Problem Solving
with
Pentominoes

Grades 1-4
Activity Book

By Alison Abrohms

Table of Contents

W9-AGB-709

Introduction

Problem Solving with Pentominoes provides hands-on puzzle activities for children in grades 1 through 4. Many of the activities can be adapted for children in the middle grades as well. The blackline masters and detailed teacher's notes have been developed to present a variety of geometric and spatial concepts. The activities encourage children to experiment with mathematical concepts and communicate their understanding of the concepts.

As children attempt to find solutions for the activities and puzzles, they will use important problem-solving strategies such as *predicting*, *deductive reasoning*, *evaluating*, *revising*, and *recording results*. The materials can be used with individual children, with pairs or small groups, or with an entire class.

The activities focus on these mathematical concepts: *transformations (slides, turns, flips)*, *tessellations (tilings)*, *symmetry*, *area*, *perimeter*, *congruence*, and *similarity*. The *More For You* section found on each activity page provides children with the opportunity to extend their learning by applying the concepts developed in the activity in a more challenging situation.

This book is designed for use with Learning Resources' *Premier Pentominoes* (LER 286) and Translucent Overhead Pentominoes (LER 415) for the overhead projector. Both sets of pentominoes are specially scored in one-inch sections. This allows young children to easily visualize area and perimeter and promotes greater understanding as children develop spatial skills. You may want to purchase additional sets of *Premier Pentominoes* to accommodate larger groups of children.

Particular attention has been given to the following standards found in NCTM's *Curriculum and Evaluation Standards for School Mathematics* (1989):

- **Mathematics and Problem Solving**
- **Mathematics as Communication**
- **Mathematics as Reasoning**
- **Mathematical Connections**
- **Geometry and Spatial Sense**

About the Book

Problem Solving with Pentominoes includes 96 pages of activity-based blackline master sheets and teacher's notes. While *Premier Pentominoes* (LER 286) and *Overhead Pentominoes* (LER 415) are available from Learning Resources, a blackline master to make paper pentominoes is provided. Teaching aids such as selected solutions, inch and quarter-inch graph paper, an award certificate, and an individual progress chart can be found at the back of the book.

Activities are organized into eight sections. Teacher's notes include the following:

Warm-Up: A manipulative activity to introduce the mathematical concepts presented in the section.

Using the Pages: An overview of each activity page.

Wrap-Up: A follow-up activity for children to demonstrate their understanding of the section concepts.

Each section also contains an activity entitled *Share at Home*. In this activity, children will be able to demonstrate their understanding of section concepts to a family member. The set of pentominoes found on pages 7 and 8 can be given to children for use at home. The children should be encouraged to return their papers to class the following day and share their solutions with their classmates.

Inch and quarter-inch graph paper is used in many lessons. Enough copies of these pages should be prepared ahead of time for children to use as needed. Children can record their solutions on these pages. The quarter-inch graph paper is suited for recording solutions to larger puzzles and may only be appropriate for older children. You may want to provide folders in which children can store their pentomino papers.

The *Selected Solutions* section provides solutions for the more challenging puzzles. For most puzzles, there is more than one correct solution. Children can record all the additional solutions they find on graph paper. If children have difficulty reaching a solution, let them work in cooperative learning groups to solve the problem. Some of the more challenging puzzles can be displayed on a bulletin board or in a math center for children to solve independently.

Throughout this book, a cooperative learning approach to problem solving is promoted using the *Think-Show-Share* strategy. Encourage children to:

- **Think** about each puzzle and a possible arrangement of pieces before they begin.

- **Show** or demonstrate their solution strategy to others.

- **Share** or discuss what they have learned.

The Pentomino Set

Pentominoes are formed by joining five congruent squares edge to edge. The name pentomino stems from penta, meaning "five," and domino, which is two congruent squares joined edge to edge. There are 12 possible geometric shapes that can be formed using five congruent squares. Each of the shapes forms a distinctive pattern often resembling a letter of the alphabet as shown:

Pentominoes can be arranged to form many interesting and recognizable shapes. As the children learn about geometric relationships by handling the pieces and fitting them together to form different shapes, their problem-solving and visual-perception skills are enhanced.

Problem Solving with Pentominoes
© 1992 Learning Resources, Inc.

Pentomino Pieces 1

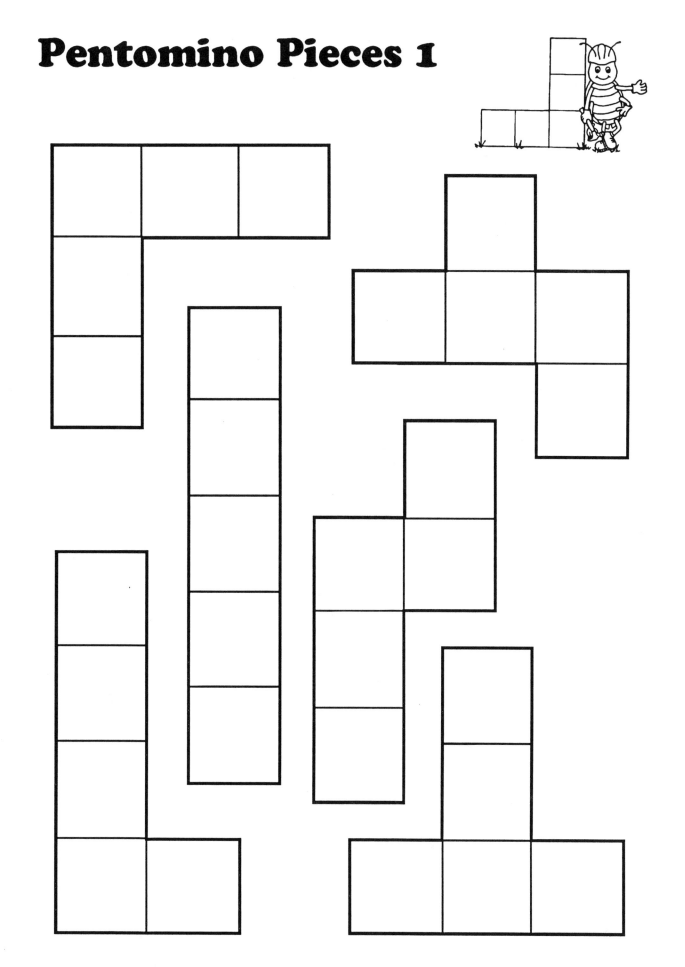

Pentomino Pieces 2

Problem Solving with Pentominoes
© 1992 Learning Resources, Inc.

Exploring Pentominoes

Teacher's Notes

The activities in this section are designed to help children become familiar with the pentomino shapes and identify any figure that does not represent a pentomino. After some exploration with congruent squares, children will develop their own definition of a pentomino.

Introduce the children to pentominoes by using the *Warm-Up* activity described below. This important introductory activity gives children the opportunity to use five congruent squares and form pentominoes through deductive reasoning and evaluation. It also encourages children to communicate mathematically as they work cooperatively with a partner. After establishing a definition of a pentomino, provide children with a prepared set of pentominoes for free exploration. During this time, let children discuss any discoveries they make about their pentominoes.

Warm-Up

Provide five one-inch squares and several copies of the inch graph paper (page 93) to pairs of children. Give children commercially made squares (Color Tiles LER 203) or squares cut from heavy paper. Then display the **T** on the overhead projector. Identify it as a pentomino and show children how it is formed from five squares. Let children model the **T** with their squares.

Show several other pentominoes on the overhead, and ask children to model each one with their squares. After children have identified and modeled several pentominoes, work together to develop a class definition of a pentomino. It may be helpful to begin by asking children what they have noticed about each pentomino they have modeled. Encourage the class to formulate these rules:

- **Each pentomino has five squares.**
- **The squares are all the same size (congruent).**
- **One side of one square fits exactly against one side of another square.**

Record the children's definition on chart paper. Then turn off the overhead projector as you arrange the following, using pentomino pieces and squares:

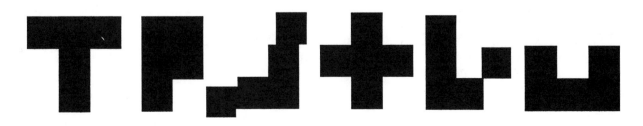

Turn on the overhead projector and ask children to identify which shape fits the pentomino definition. Once children understand the definition of a pentomino, ask them to use their squares to see how many different pentominoes they can make. Each pentomino they identify can be recorded and cut from the inch graph paper.

At the end of the session, compile a class set by pasting one model of each pentomino identified on the chart paper below the definition. During this process, if children offer a pentomino that is the same as one already on the chart paper, help them flip or turn the model to find the one it matches.

Some children may present a model that does not fit the definition of a pentomino. You may want to discuss these examples as a class and decide why they do not fit the definition. However, praise the children for their efforts. Help children model any of the pentominoes that have not been identified. (There are 12 possible pentominoes.) Display a model of each pentomino on the chart paper.

Free Exploration
After children have had the opportunity to model and form the pentominoes with the inch squares, distribute a pentomino set to each child or pair of children. The scoring of each inch section in the set of *Premier Pentominoes* is particularly effective in helping children make the transition from shapes formed by the five squares to the pentomino pieces.

Encourage children to manipulate the pentominoes without any instruction. This free exploration will allow them to begin thinking about the relationships among the pieces and how they fit together. Observe the children as they work and question them about their observations. For example, some children may recognize that each piece resembles a letter of the alphabet, while others may recognize that some pieces are symmetrical (**U, X, V, T, W, I**).

 **Using the Pages**

Page 12
Pentomino Shapes: Give each child or pair of children a pentomino set. If additional sets are needed, duplicate pages 7 and 8 onto heavy paper, and have children carefully cut apart the pieces. Display the pentominoes one at a time on the overhead projector, and ask children to hold up the corresponding pieces from their sets.

Distribute page 12. Ask children to find the pentomino pieces that cover the shapes and place them on the page. To help younger children find these pieces within the set, hold up each of the pictured pieces one at a time. Ask children to find the piece, then place it on the page.

Page 13
More Pentomino Shapes: Distribute page 13 and continue the matching activity using the shapes on the page. Then have children look at the first shape and ask them to identify the capital letter the shape looks like. Some children may readily see the capital

letter the pentomino resembles. For children who have difficulty visualizing the resemblance, print a capital letter on acetate and place it next to the corresponding pentomino piece on the overhead projector.

For example:

When children are confident in recognizing the pentomino pieces as letters, name a letter and ask students to display the corresponding pentomino.

Page 14
Find the Pentominoes: This activity page requires children to identify pentomino shapes. First, arrange five squares on the overhead projector into various geometric shapes. Include shapes that fit the pentomino rules and shapes that do not. Encourage children to identify whether each shape is a pentomino and discuss why the shape does or does not fit the pentomino rules. Then have the children complete page 14.

Page 15
Find More Pentominoes: Review the pentomino shapes by asking children to recall the rules for identifying a pentomino (each pentomino has five squares, the squares are all the same size, one side of one square fits exactly against one side of another). Then ask children to complete page 15.

Page 16
Share at Home: Give each child a copy of page 16, the letter to the parents, and copies of pages 7 and 8. Explain to children how the *Share at Home* activity is to be used. Read the instructions and tell them that they are to use the cutouts from pages 7 and 8 to identify which shapes on the page are pentominoes for family members. You may also want to send home five one-inch squares with children so they can model shapes that are not pentominoes and discuss the pentomino rules.

Wrap-Up
Distribute pentomino sets to each child or pair of children. Arrange several pentomino pieces on the overhead projector while it is turned off. Include other shapes formed by five squares that do not fit the pentomino rules. Turn on the projector. Let volunteers use their pentomino pieces to help them identify which of the shapes on the projector are pentominoes.
Ask volunteers to describe why shapes that are not pentominoes do not fit the rules.

Pentomino Shapes

Name _____

These are some of the shapes in your set of pentominoes.
Cover the shapes with your pentomino pieces.

More For You
Work with a friend:
• Hold up any pentomino piece from one set.
• Can your friend hold up the matching piece from another set?
• Try again with another piece.

Problem Solving with Pentominoes
© 1992 Learning Resources, Inc.

More Pentomino Shapes

Name_____

Cover the shapes with your pentomino pieces.

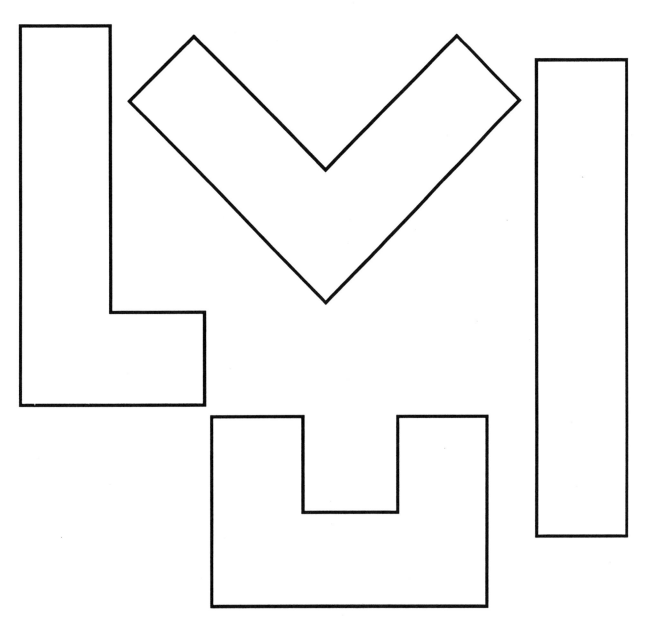

More For You
- Choose any pentomino piece.
- Trace around it on the back of this paper.
- Ask a friend to find a piece that covers the shape.
- Try again with another piece.

Problem Solving with Pentominoes
© 1992 Learning Resources, Inc.

Find the Pentominoes

Name _____

Try to cover these shapes with your pentomino pieces.
Circle the shapes that are pentominoes.

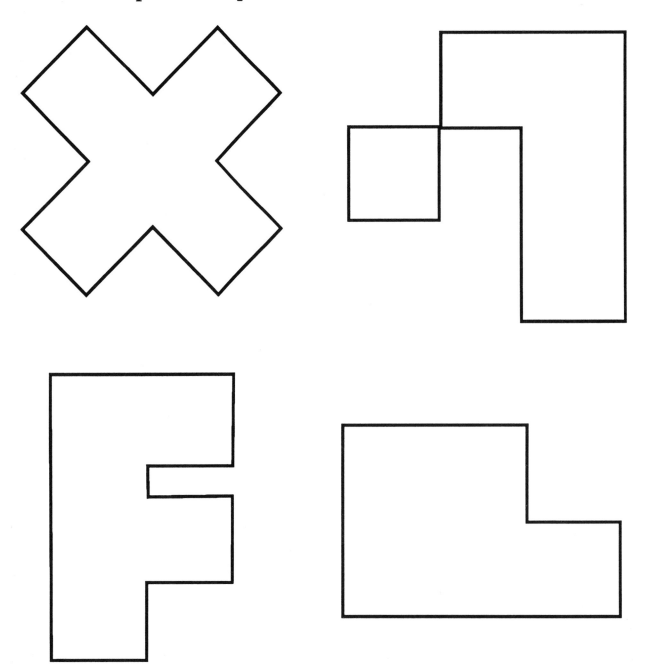

More For You
- Look at any shapes you did not circle.
- Can you tell why this shape is **not** a pentomino?

Problem Solving with Pentominoes
© 1992 Learning Resources, Inc.

Find More Pentominoes

Name_____

Try to cover these shapes with your pentomino pieces.
Circle the shapes that are pentominoes.

More For You
• Fit these pieces together to make a shape.

• Trace around the shape on the back of this paper.
• Draw a picture in your shape.

Share at Home

Name _____

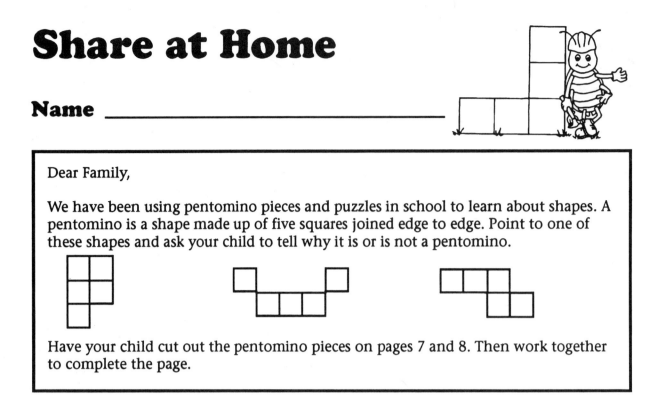

Dear Family,

We have been using pentomino pieces and puzzles in school to learn about shapes. A pentomino is a shape made up of five squares joined edge to edge. Point to one of these shapes and ask your child to tell why it is or is not a pentomino.

Have your child cut out the pentomino pieces on pages 7 and 8. Then work together to complete the page.

Compare your pentomino pieces with these shapes.
Circle any shape that is a pentomino.
Tell why the other shapes are not pentominoes.

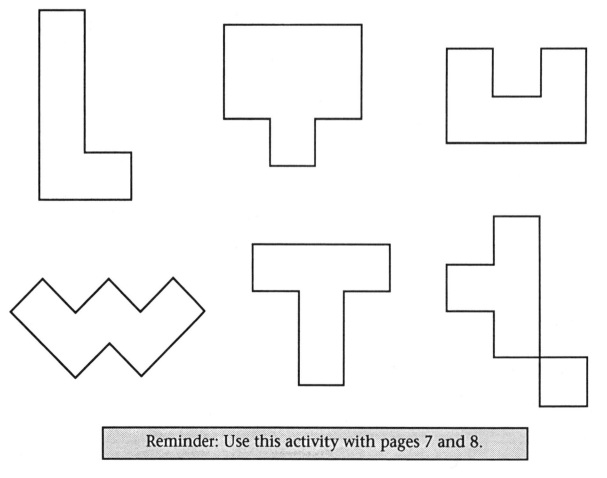

Reminder: Use this activity with pages 7 and 8.

Problem Solving with Pentominoes
© 1992 Learning Resources, Inc.

Moves & Patterns

Teacher's Notes

This section introduces children to three kinds of motion in geometry—*translations (slides), rotations (turns), and reflections (flips)*. With this series of activities, children should realize that after sliding, flipping, or turning a pentomino piece, the shape of the piece remains the same; only its position changes.

Children will also explore tessellations, or tilings, in this section. Nine pentominoes will produce tessellating patterns with no gaps: **I, L, N, P, V, W, X, Y, Z**. The remaining three pieces—**F, T,** and **U**—will also tile a surface, although these pieces will need to be turned upside down to produce a tessellation without gaps. For interested children, you may want to present some work by artist M. C. Escher (1898-1972) who created fascinating designs using tessellating patterns.

Warm-Up

Use this demonstration activity to give children the opportunity to see how motion affects three-dimensional objects. On a table, display large three-dimensional objects that will stand upright such as large, thick books or boxes. Each object should have identifiable features on several of its sides.

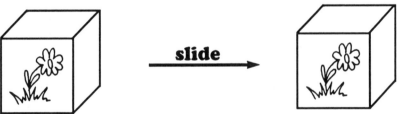

To begin, slide the objects around the table and ask children to note what types of changes occur. (Children should recognize that the objects remain the same but are in different locations on the table.)

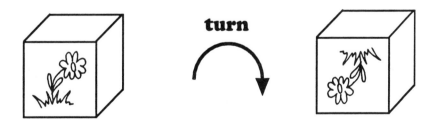

Next, turn one of the objects 180° and ask children to describe the move. (Accept any reasonable description.) Continue to demonstrate other turns such as a 90° turn and a 270° turn.

Finally, flip several of the objects from front to back. Children should recognize that this motion is different from a turn. Help children agree on a term that describes this motion. (Children should be able to describe how the location or position of the object changes, yet realize the object's shape remains constant.)

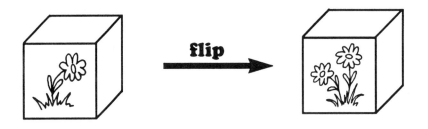

flip

Using the Pages

Page 21

Slides: Distribute pentomino pieces to pairs of children. Display a pentomino piece on the overhead projector and have children use the same pentomino and follow along as you demonstrate a "slide." Have children slide the piece *up*, *down*, *right*, and *left*. Continue the demonstration with other pieces. Then provide children with page 21 and let them complete the activity independently.

Page 22

Turns: As with page 21, begin this activity with a demonstration on the overhead projector as children follow along with their set of pentominoes. Show children how to rotate pieces 90° (quarter turn) and 180° (half turn). Depending on the age of the children, you might want them to become familiar with these terms: *90° turn* or *quarter turn*, and *180° turn* or *half turn*.

Distribute page 22 and read the directions with children. Then let them work independently. After completing this activity, children should notice that the shape of each pentomino remains the same after a turn; only its location is different.

Page 23

Slide or Turn? Use the **U** from two sets of pentominoes to demonstrate a slide and a turn on the overhead projector. Turn off the overhead projector as you arrange the pieces in the *before* and *after* positions to demonstrate each of the motions. Turn on the overhead projector and ask children to describe the motion. Continue with other demonstrations. Then give children page 23 to complete independently.

Page 24

Flips: Demonstrate the flipping motion on the overhead projector as children follow along using their pentomino pieces. Trace around several pieces as you flip them to illustrate *before* and *after* positions. If possible, provide children with small mirrors to let them discover that a flip is the same as a reflection. Children may also discover that, in some cases, a flip is the same as a 180° or 270° turn. As with a slide and a turn, the shape of any pentomino piece remains the same after a flip. (For example, an **I** remains an **I** after it has been flipped, turned, or slid.)

Have children complete page 24. Then point out that some pieces will look the same whether they are flipped over or whether they are turned a half turn. Ask children to use inch graph paper (page 93) to practice these moves with some pentomino pieces. Help them make a list of the pentomino pieces that stay the same after a flip (**I, T, U, W, X**).

Page 25
Which Piece Matches? Children can work independently or with partners to complete this page. Give children pentomino pieces and ask them to model each motion with their pieces before choosing the correct answer. The slide, turn, and flip are all used in this activity.

Page 26
Cover Some Shapes: This activity requires children to turn or flip their pentomino pieces to make them fit within the shape outlines. Before distributing the activity page, ask children to practice turning and flipping some of the pentomino pieces on their desks. Then let them complete page 26 independently.

Page 27
Sliding Patterns: Interesting tessellating (tiling) patterns can be made by sliding, turning, or flipping the pentomino pieces. Children can experiment to find various tessellations. Then let them complete page 27. Encourage children to shade or decorate patterns to produce intricate designs. For example:

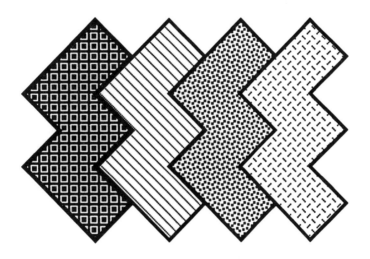

Point out any tessellating patterns found in the classroom or school such as those sometimes found in floor or wall tiling. Then provide children with inch graph paper (page 93) on which they can trace some tessellating patterns.

Page 28
Turning Patterns: Place the **L** pentomino piece on the overhead projector and trace around it. Then turn and flip the piece to make the tessellating pattern shown. Have children describe how you moved the pentomino as you made the pattern. Ask volunteers to add to the pattern by moving the piece to new locations. Children may wish to make the

pattern on their own paper and decorate each pentomino with brightly colored designs. Then distribute page 28 and let children complete the page independently.

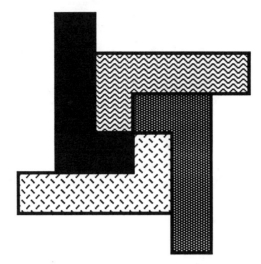

Page 29
Share at Home: Give each child copies of pages 7, 8, and 29. Read the letter and review the content so that children are familiar with the activity and can easily demonstrate slides, flips, and turns for family members.

Wrap-Up

After completing the activities in this section, children should be able to determine if a motion is a slide, a turn, or a flip. With the overhead projector off, trace around each piece, one at a time, to display the outlines to the right. Turn on the projector and ask a volunteer to place the pentomino in each of the outlines and then tell which motion would describe the movement (**Y**, slide; **L**, turn; **P**, flip).

Pairs of children can continue this activity by tracing around other pentomino pieces to record a slide, a flip, or a turn and then challenge each other to identify the motion.

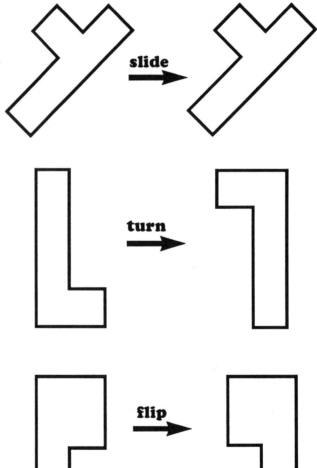

slide

turn

flip

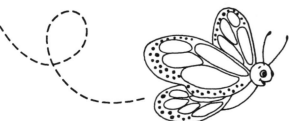

Problem Solving with Pentominoes
© 1992 Learning Resources, Inc.

Slides

Name _____

Slide each pentomino piece.
Trace around it.

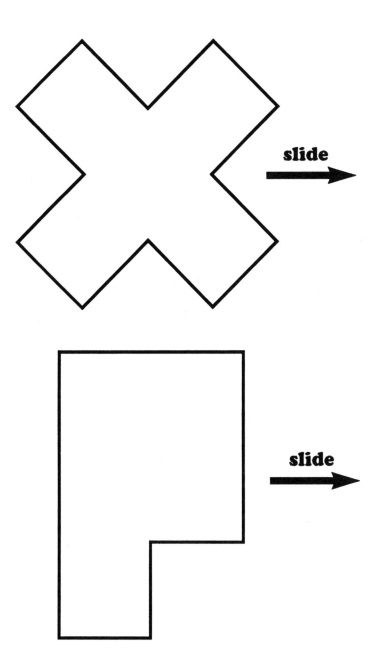

slide

slide

More For You
• Slide other pentomino pieces.
• How do the pieces look after you slide them?

Turns

Name _____

Turn each pentomino piece.
Trace around it.

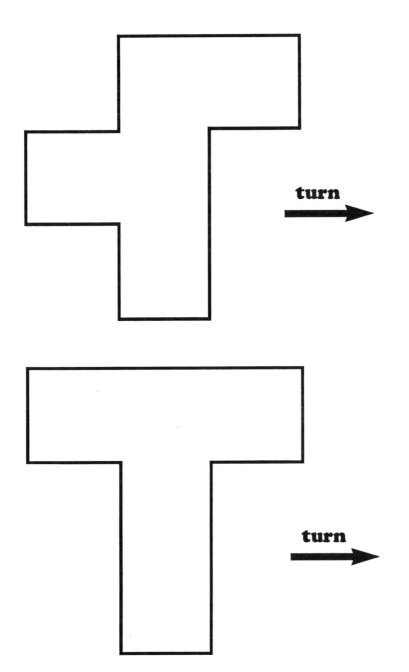

turn →

turn →

More For You
• Turn other pentomino pieces.
• How do the pieces look after you turn them?

Problem Solving with Pentominoes
© 1992 Learning Resources, Inc.

Slide or Turn?

Name _____

Use your pentomino pieces.
Is it a slide or a turn?
Circle your answer.

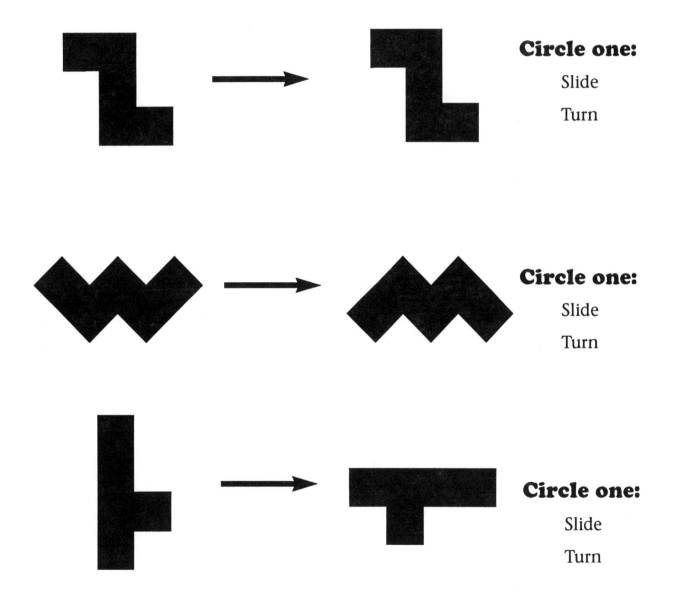

Circle one:

Slide

Turn

Circle one:

Slide

Turn

Circle one:

Slide

Turn

More For You
• Trace around a pentomino piece on the back of this paper.
• Slide it or turn it, then trace around it.
• Ask a friend to tell if it is a slide or a turn.

Flips

Name _____

Find each pentomino piece.
Flip it and trace around it.

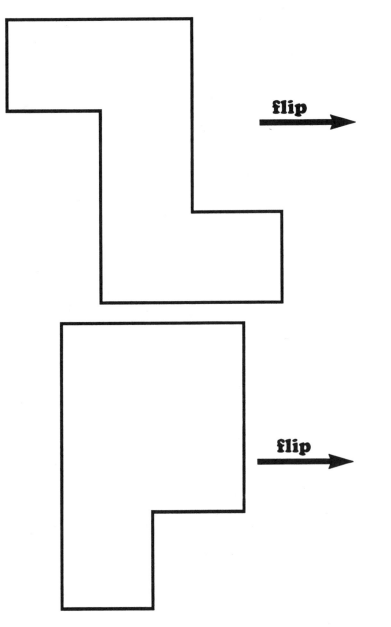

flip →

flip →

More For You

- Find the **W** piece and trace around it on the back of this paper.
- Predict how it will look if you flip it down.
- Flip it down and trace it. What do you notice?

Problem Solving with Pentominoes
© 1992 Learning Resources, Inc.

Which Piece Matches?

Name _____

Circle the piece in each row that matches the
pentomino in the left hand column.
Use your pentomino pieces.
Flip or turn to find the match.

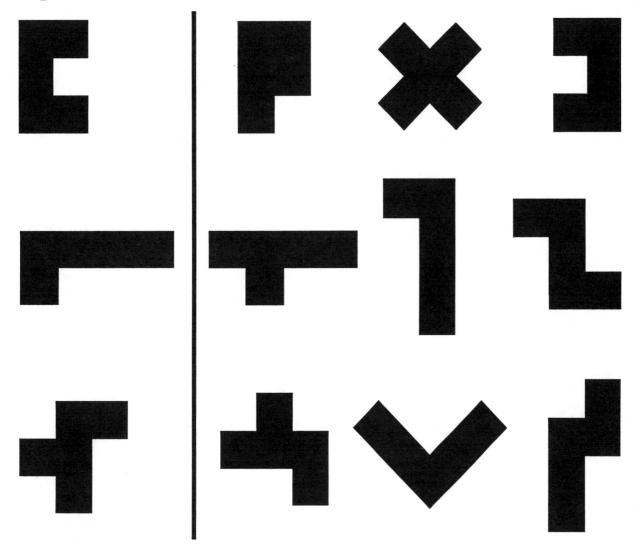

More For You

• Find the ◤◢ and trace around it on the back of this paper.
• Put two pieces next to the one you traced.
• Use the ◤◢ as one of the pieces. Flip or turn the pieces.
• Ask a friend to tell which piece matches the one you traced.
• Try the activity again with other pieces.

Cover Some Shapes

Use these pentomino pieces: **L ⊥ P**

Slide, turn, or flip the pieces to cover each shape.
Trace around each piece.

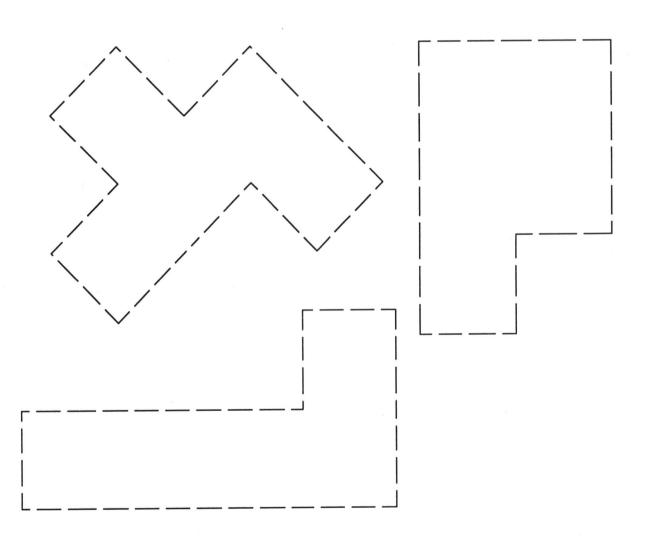

More For You
• Trace around other pentomino pieces on the back of this paper.
• Ask a friend to cover each shape with a pentomino piece.

Sliding Patterns

Name _____

You can slide the  piece to make a design without gaps.
Put your pentomino piece on the design.
Trace and Slide.
Color your design.

More For You
• Use another sheet of paper and the piece.
• Trace around the piece. Slide it and trace again.
• Make a design without any gaps.
• Color your design.

Problem Solving with Pentominoes
© 1992 Learning Resources, Inc.

Turning Patterns

Name _____

You can slide the **L** to make a design without gaps.
Put your pentomino piece on the design.
Trace and turn.
Color your design.

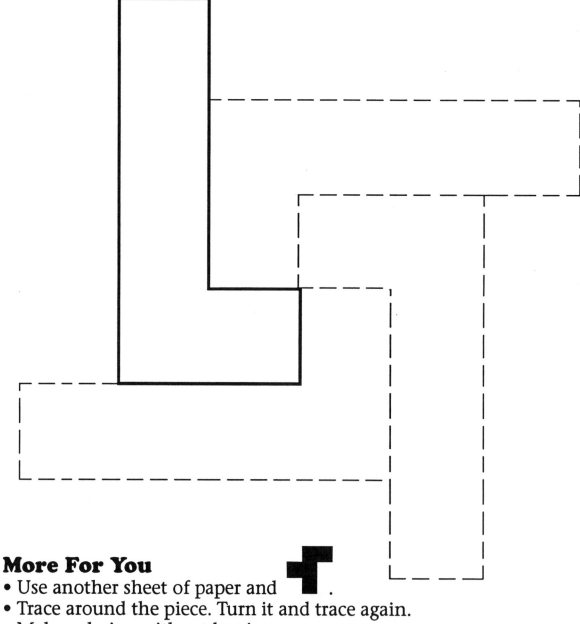

More For You
- Use another sheet of paper and ▟.
- Trace around the piece. Turn it and trace again.
- Make a design without leaving any gaps.
- Color your design.

Problem Solving with Pentominoes
© 1992 Learning Resources, Inc.

Share at Home

Name _____

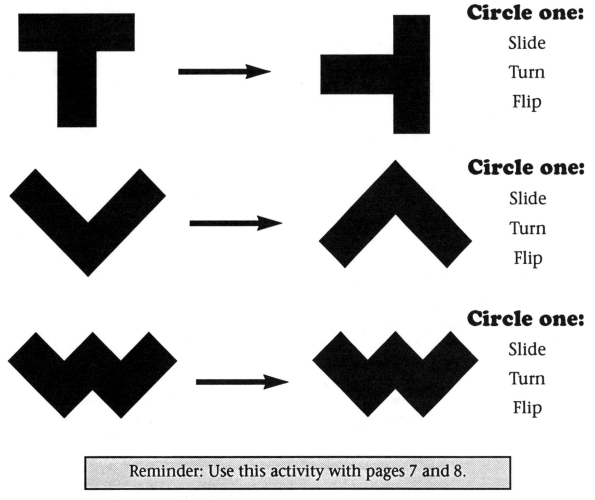

Show each move with your pentomino pieces.
Tell if it is a slide, turn, or flip.

Circle one:

Slide

Turn

Flip

Circle one:

Slide

Turn

Flip

Circle one:

Slide

Turn

Flip

Reminder: Use this activity with pages 7 and 8.

Covering Geometric Shapes

Teacher's Notes

With the activities in this section, children will begin placing pentomino pieces on shapes comprised of two or more pieces. The activities move strategically from shapes made with two pieces with all interior lines shown, to shapes made with four or more pieces with only an outline given.

As the shapes increase in difficulty, children will need to begin visualizing how they could best position specified pentomino pieces to cover a shape. The activities at the end of this section focus on determining how to place pentomino pieces within shapes. Because of the configurations of the pentomino pieces, many shapes can be covered in more than one way. Encourage children to look for alternative ways to cover each shape.

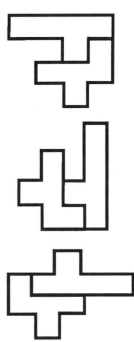

Warm-Up

Use the **F** and **Y** pieces to make the shapes shown. Trace around each shape on a sheet of acetate and display on the overhead projector. Place the **F** in the upright position below the first shape and ask, **What can I do to make the F fit in the shape?** Children should draw from their experiences in the previous section and suggest turning the piece to make it fit. Similarly, place the **Y** in an upright position below the first shape and let volunteers model how the piece can be flipped to fit the shape. Follow the same procedure to cover the remaining shapes.

Using the Pages

Page 33
Covering Shapes With Two Pieces: In this activity, children will use specified pentomino pieces to cover the outlines. On the activity page, the shapes and interior lines are given to assist children. Each shape can be covered with the same pieces. This will help children realize that by turning or flipping pentomino pieces, other shapes can be formed. As children complete the *More For You* section, you may want to have them trace their shapes on a clean sheet of paper, then bind these pages into a classroom book of pentomino puzzles.

Page 34

Covering Shapes With Three Pieces: Use the **L** and **N** pentomino pieces on the overhead projector to review the two forms from the previous lesson. Remind children how the pieces are turned or flipped to make the new shapes. Then read the directions for page 34 and point out to children that they will use three pentomino pieces to cover each shape on the page. Then let them work independently.

Page 35

Covering Shapes With Four Pieces: Four pentomino pieces are used to cover this shape. As in the previous activities, the pentomino pieces are specified, and interior lines are shown within the shape. In the *More For You* section, children will be asked to cover the same outline with a different set of pieces. If children experience difficulty with this task because of the interior lines, you may want to trace only the outline of the shape on a clean sheet of paper.

Page 36

Numbers: Ask children to select the pentomino pieces that will form the number 4. Remind them that they may need to turn the pieces to find the ones to make the number. Then let children work independently to form the numbers 1, 6, and 9. Have them use inch graph paper (page 93) to record the pieces they used to make each number.

Page 37

A Shape to Cover: Have children cover the shape with the pentomino pieces shown by the interior lines. Encourage children to turn or flip the pieces if they have difficulty visualizing how each piece will fit in the shape.

Page 38

Where Do the Pieces Fit? In this activity, children will complete a partially covered shape without the benefit of interior lines. After children place the **L** on the shape, ask, **Which pentomino piece should you try on the shape next? Why?** As children examine the remaining space, direct the discussion so children realize that the **V** is best placed next on the shape as it has only one possible placement. Once the **V** is on the shape, it will become apparent where the **N** will fit. This 3" x 5" rectangle can be covered with seven different groups of pentomino pieces. (See Selected Solutions.)  Provide children interested in finding all the possible combinations with several pieces of inch graph paper (page 93).

Page 39

Turn or Flip to Fit: As children continue to work through the activity pages in this section, their ability to manipulate pieces to cover a given shape will improve. This activity focuses on turning and flipping pieces to cover a given shape. After children have placed one piece on the puzzle, ask them to think about how they must turn or flip the remaining pieces to make them fit.

Page 40

Place the First Piece: Trace the shape shown on page 40 on a sheet of acetate and place it on the overhead projector. Place the **F** as shown on the activity page. Ask: **Is this a good place to put this piece? Why or why not?** At this point, children should be able to respond that this placement is not reasonable because the remaining spaces cannot be covered with the other pieces. Place the piece on the acetate incorrectly again and let children consider why this placement is also incorrect. Then let children complete the page.

Page 41
More Space to Cover: In this activity, all of the pentomino pieces needed to cover the shape are not specified. Children must find the pieces from their set to cover the shaded portion of the shape, then they can choose two of the three pieces shown to cover the remaining space. Encourage children to look at the space that is left after they place a piece to determine which piece will cover the remaining space. When the activity is completed, let children compare their solutions.

Page 42
Which Pieces? Before they begin, ask children whether they "see" any pentomino pieces in the shape. For example, children may notice the **T** shape on the left side of the puzzle. Remind children that the shaded part in the shape does not get covered. Then let children work independently or in small groups to complete the activity.

Page 43
Share at Home: Give each child copies of pages 7, 8, and 43. Remind children to carefully look at the space they will cover to "see" if they can predict which set of pieces will cover the shape.

Wrap-Up
Form the shape shown with the **P**, **X**, and **T** pieces. Trace around the shape on acetate and display it on the overhead projector along with these pieces: **I, L, P, X, T**. Ask children to examine the shape carefully and predict which pieces will cover it. Let volunteers take turns placing pieces on the shape until it is covered.

Then give paper and a set of pentominoes to pairs of children. Ask each pair to form a shape with four pieces and trace around the outline of the shape. Then have pairs exchange papers and solve each other's puzzles.

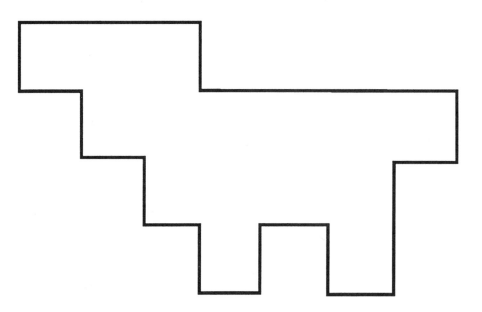

Covering Shapes With Two Pieces

Name _____

Use these pieces to cover each shape.

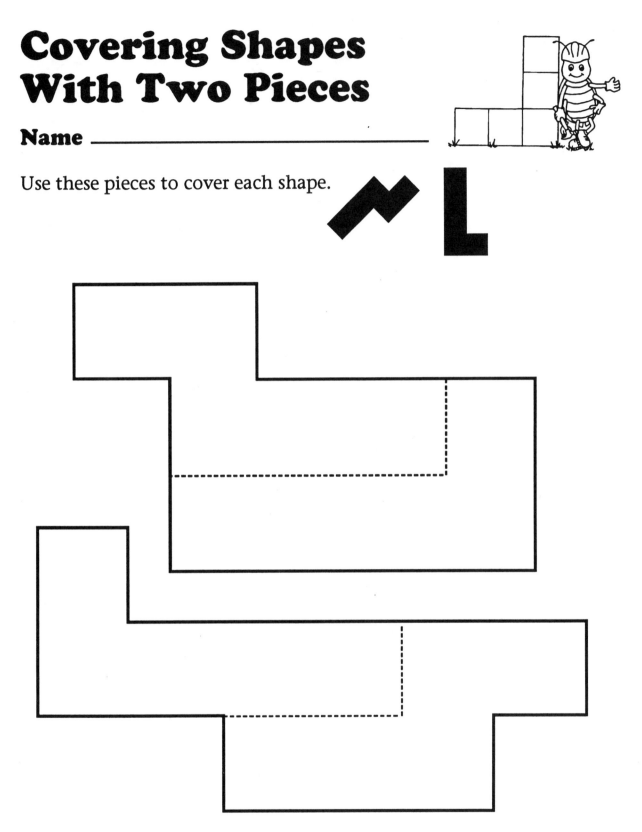

More For You
- Use two other pentomino pieces.
- Make a shape on the back of this paper.
- Trace around each piece.
- Color your shapes.

Problem Solving with Pentominoes
© 1992 Learning Resources, Inc.

Covering Shapes With Three Pieces

Name _____

Use these pieces to cover each shape.

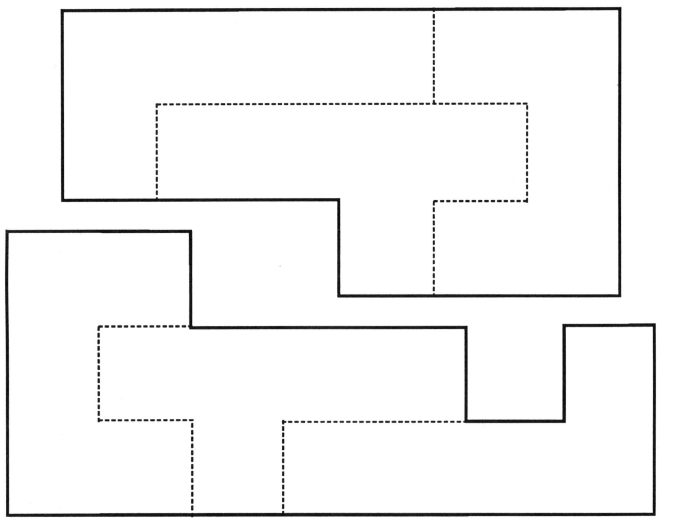

More For You
• Use three other pentomino pieces.
• Make shapes on the back of this paper.
• Trace around each shape.
• Ask a friend to cover them with pentomino pieces.

Problem Solving with Pentominoes
© 1992 Learning Resources, Inc.

Covering Shapes With Four Pieces

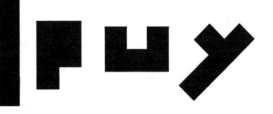

Name _____

Use these pieces to cover each shape.

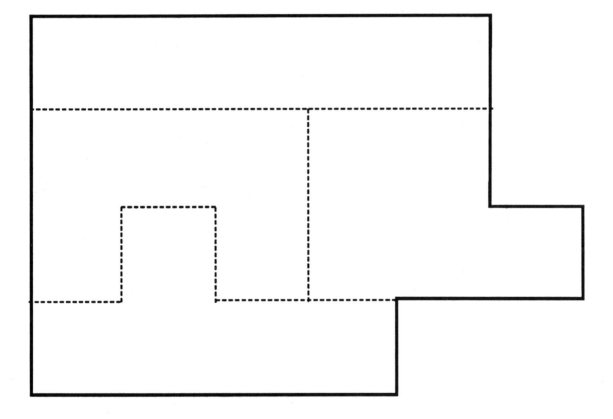

More For You
• Use these pentomino pieces to cover the shape.

• Slide page 93 (inch graph paper) under the pentomino pieces.
• Trace around them.

Numbers

Name _____

Use the pentomino pieces to cover the number.

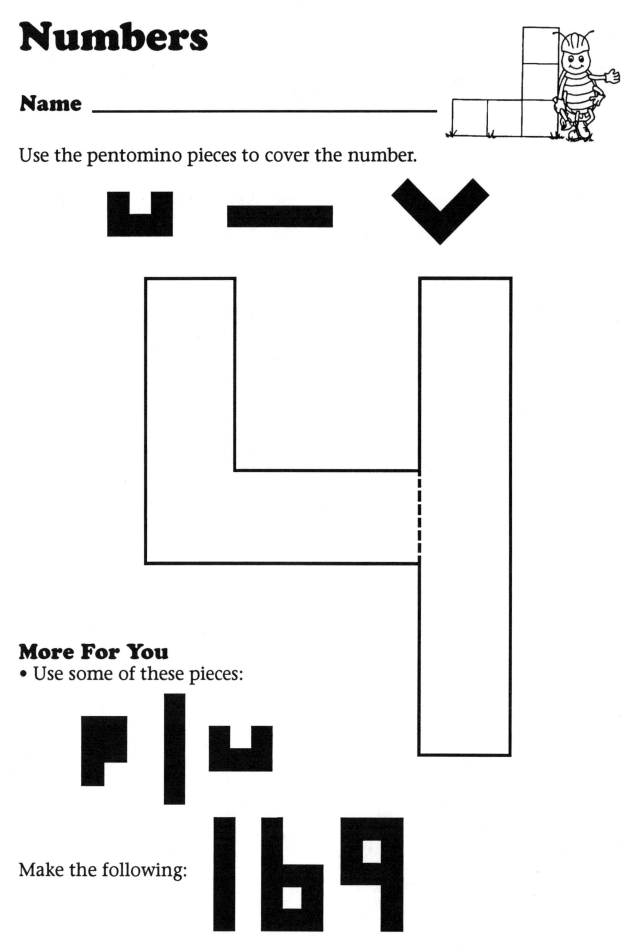

More For You
• Use some of these pieces:

Make the following:

Problem Solving with Pentominoes
© 1992 Learning Resources, Inc.

A Shape to Cover

Name _____

Use these pentomino pieces to cover the shape.

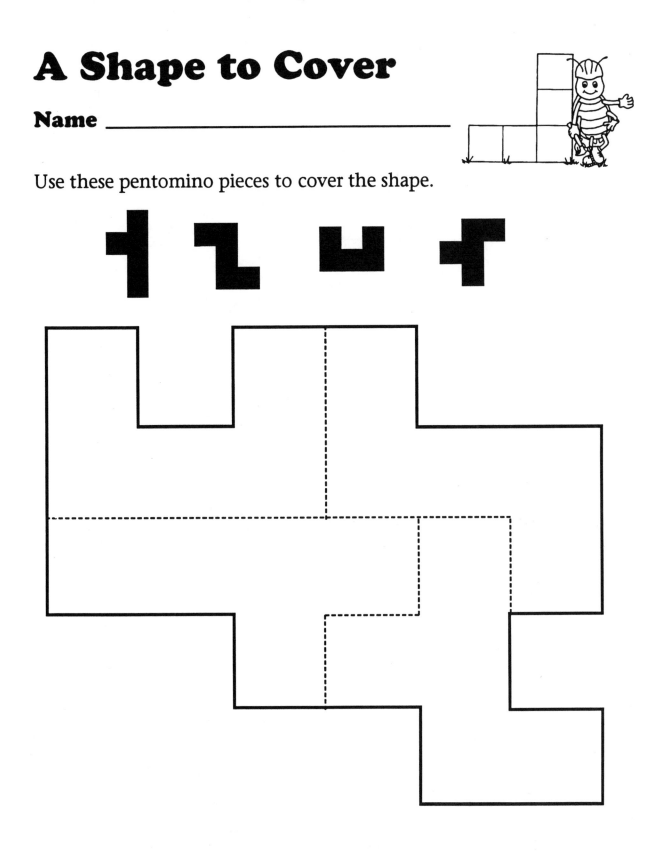

More For You
• Make another shape with the same pieces.
• Trace around each piece on the back of this paper.
• Ask a friend to cover the shape with pentomino pieces.

Where Do These Pieces Fit?

Name _____

These pieces fit on the shape.

You can see where the **L** fits.
Slide, turn, or flip the other pieces to cover the shape.
Trace around each piece.

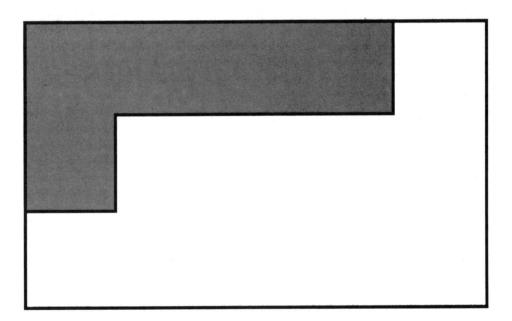

Can you cover the shape with these pieces?

More For You
• Work with a friend.
• Find other pieces to cover the shape.

Problem Solving with Pentominoes
© 1992 Learning Resources, Inc.

Turn or Flip to Fit

Name _____

Hold these pieces like this.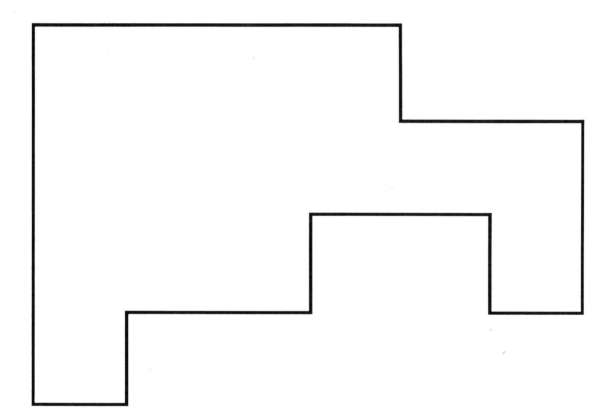

Turn or flip them to fit the shape below.
Trace around each piece.

More For You
• Cover the shape using these pieces:

Place the First Piece

Name _____

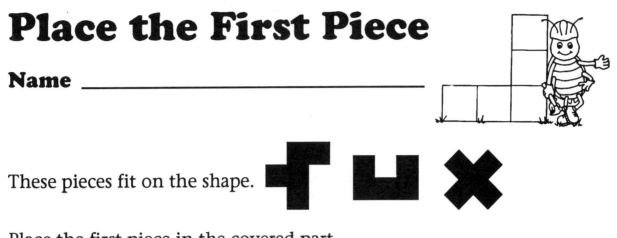

These pieces fit on the shape.

Place the first piece in the covered part.
Is it a good place for it?

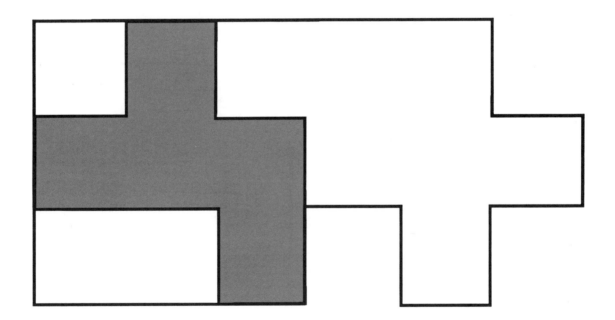

Now cover the shape.

More For You
• Can you cover the shape another way?
• Trace the pieces you used on the back of this page.

Problem Solving with Pentominoes
© 1992 Learning Resources, Inc.

More Space to Cover

Name _____

Part of the shape is covered.
Place your pentomino pieces on the covered part.
Choose two of these pieces to cover the rest of the shape.

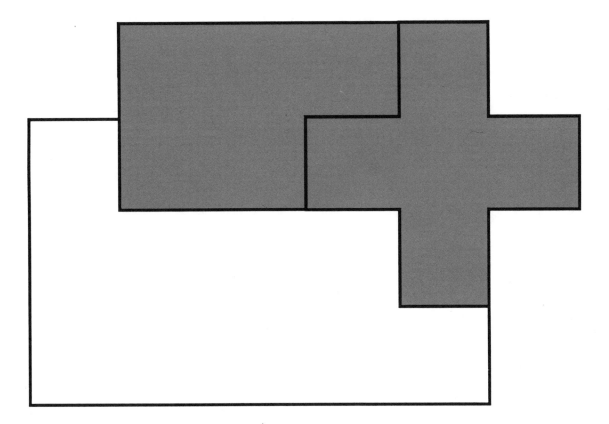

More For You
- Trace around the pieces you used to cover the rest of the shape.
- Share it with a friend.
- Did you both cover the shape the same way?

Problem Solving with Pentominoes
© 1992 Learning Resources, Inc.

Which Pieces?

Name _____

Three of these pieces fit on the shape.

Look carefully at the shape to predict which pieces you will need.
Trace around each piece you use.
Do not cover the shaded part.

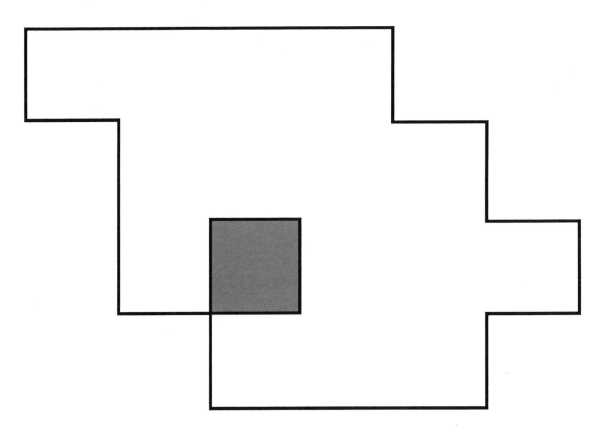

More For You

•Could you tell the ▮ would not fit? How? Talk about it.

Problem Solving with Pentominoes
© 1992 Learning Resources, Inc.

Share at Home

Name _____

Dear Family,

We have been using pentomino pieces in school to cover shapes. Have your child cut out the pentomino pieces on pages 7 and 8. Then ask the child to predict which set of pieces would cover the shape and circle the pieces. Then cover the shape with those pieces.

Predict which set of pentomino pieces will cover the shape.
Circle the pieces. Try it.

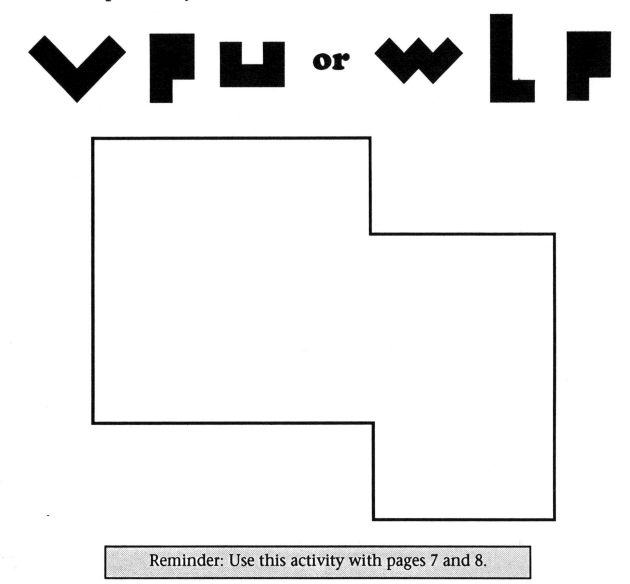

Reminder: Use this activity with pages 7 and 8.

Puzzles & Games

Teacher's Notes

In this section, children will focus on covering geometric shapes using some of their pentomino pieces. The problems are more challenging than those on previous pages as interior lines are not given, and particular pentomino pieces are not specified. Rather, the number of pentomino pieces needed to cover a shape will be given. To solve the problems, children will need to look carefully at the shape they want to cover and notice any characteristics. In the shapes below, it is possible that the **W** or the **F** could cover the angled space on the right. The visual clues will help children as they choose pieces to try on the shape.

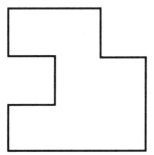

(Solutions: **W, P** and **F, U**)

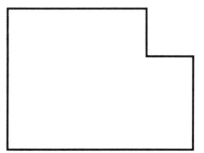

(Solutions: **F, Y, U** and **W, P, Y**)

To cover these shapes, children can use all the problem-solving strategies introduced in the previous sections. Some of these strategies include the following:

- **Using motion—slide, turn, or flip—to manipulate the pentomino pieces;**
- **Determining which pieces should be placed initially on a shape;**
- **Deciding which placements are reasonable;**
- **Examining the remaining space in a shape;**
- **Predicting which pieces will not fit in a shape.**

Encourage children to use these strategies as they complete the problems in this section. Working with a partner or in small groups will enhance learning as children verbalize their thinking strategies and communicate their solutions.

Warm-Up

In this section, children will be asked to find more than one set of pentomino pieces to cover a shape. Construct the shapes shown on page 47 and trace the outside of each shape on a sheet of acetate. Display a set of pentomino pieces and one of the shapes on the overhead projector. Let volunteers find both ways to cover each shape. Prompt children to think carefully about each piece by asking them to tell why they think the piece will fit in the shape.

Using the Pages

Depending on the age of the children, provide either inch graph paper (page 93) or quarter-inch graph paper (page 94) for them to record their solutions. On the quarter-inch graph paper, children can either trace their pentomino solutions to make a full-size model or shade squares to make a smaller model. If the quarter-inch graph paper is used, let children practice recording some of the pentomino pieces so that they understand how to count the squares to draw a pentomino.

Page 47

Two of the Same: Encourage children to look carefully at the shapes. Then ask, **Would the I piece fit on the shape? Why or why not?** (No, it is too long.) Children can move pieces that will not fit on the puzzle to a separate pile. By narrowing the number of pieces they have to work with, they will be able to find the solution more easily. Let children work independently or in small groups to complete the page.

Page 48

A Shape Puzzle: Read the directions with children and let them complete the activity. Point out that there are at least seven different ways to cover this shape. Have children record their solutions using the inch or quarter-inch graph paper. Then compile a class list of each solution identified.

Page 49

Cover the Stairs: Help children read the directions. Point out to children that there are several solutions to the puzzle and that they are to use separate sets of pentomino pieces to cover each set of stairs. After children cover each shape, ask, **How could we tell if both shapes are exactly the same?** If children do not suggest it, ask them to place one set of pentomino pieces on top of the other, and turn to match.

Page 50

Making Rectangles: There are seven possible ways to cover this 3"x 5" rectangle with one set of pentomino pieces. Give children copies of page 93 or 94 on which they can record their solutions.

Page 51

Making Large Rectangles: Depending on the age and interest level of the children, you may want to encourage them to work independently to find the solutions. If children are unable to draw their pentomino arrangements on the graph paper provided, distribute inch graph paper (page 93) on which they can trace their solutions. There are 25 possible ways to cover this 4"x 5" rectangle with one set of pentomino pieces.

Page 52

A Dog: Ask children to identify the shape and discuss why it resembles a dog. Then have them choose pentomino pieces to cover the shape. Remind them to slide, flip, and turn pieces as they work until they discover a set of five pieces to cover the shape. If children have difficulty, work together as a class using the overhead projector to experiment with the various pieces until the solution is found.

Page 53

A Butterfly: Trace the shape on acetate and display it on the overhead projector. Have children select pentomino pieces to cover the shape and tell why each piece fits in a particular location. Then let children work independently to cover the shape.

Page 54

A Flag: Encourage children to use visual clues to identify some pieces that might fit (the V in a corner or an I on the left side). Children can trace around each piece to record their solutions, or cut out the pentomino pieces from pages 7 and 8 and paste these pieces on the shape.

Page 55

Making Squares: There are six possible ways to cover this five-inch square with one set of pentomino pieces. Since the number of pieces is not given, you might want to pose this situation: **If we place this square on the inch graph paper, we will see twenty-five squares within our square. How many pentomino pieces will be needed to cover this square?** (five) Children can record their solutions on graph paper (page 93 or 94).

Page 56

A Square Game: Provide page 56 and one set of pentominoes for each pair of children. Children should understand that pieces cannot overlap or extend beyond the game board outline. As they play, children might realize that longer pieces placed in the middle of the game board will effectively reduce the number of pieces that will fit and may block their opponent from placing another piece. To challenge children, pose this problem: **Can this square be completely covered with pentomino pieces? Why or why not?** Children may recognize that since this is a 6"x 6" square (area of 36), it is not a multiple of 5 and cannot be covered with pentomino pieces.

Page 57

Share at Home: Give each child copies of page 7, 8, and 57. Read the instructions and tell children to use the pentomino set to complete the activity at home with their family. You may also want to give children a copy of page 56, *A Square Game,* to play at home.

Wrap-Up

Trace the outline of the flag on page 54 on a sheet of acetate and display it on the overhead projector. Ask for volunteers to place pentomino pieces on the shape and explain their strategy for each placement.

Problem Solving with Pentominoes
© 1992 Learning Resources, Inc.

Two of the Same

Name _____

Cover each shape with 2 pentomino pieces.
Trace around the pieces to show the answer.

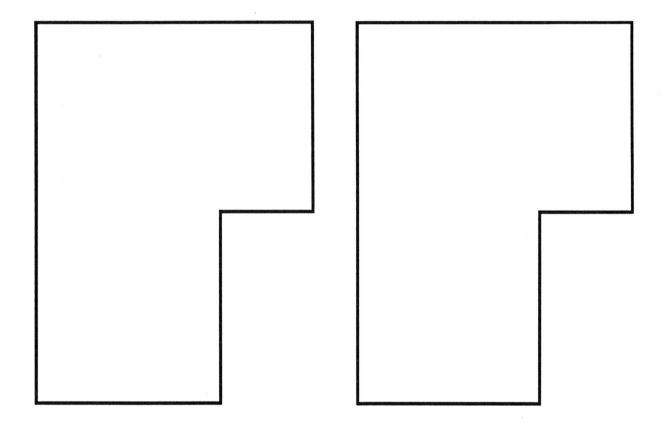

More For You
• Find another way to cover the shape.
• Trace around the pieces on the back of this paper.
• Did you find the same ways to cover the shape as a friend did?

A Shape Puzzle

Name _____

Cover the shape with 3 pentomino pieces.
Trace around the pieces to show the answer.

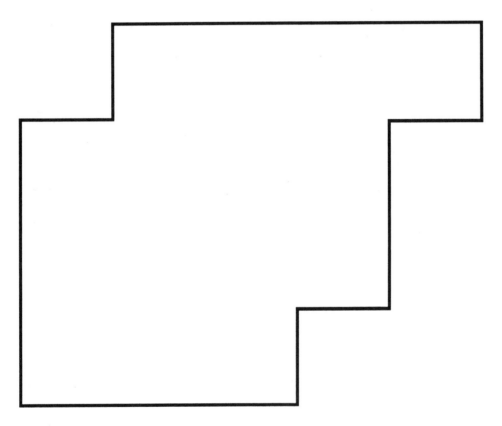

More For You
- Find other ways to cover the shape.
- Trace around the pieces on the graph paper.

Problem Solving with Pentominoes
© 1992 Learning Resources, Inc.

Cover the Stairs

Name _____

Cover both sets of stairs with 3 pentomino pieces.
Trace around the pieces to show the answer.

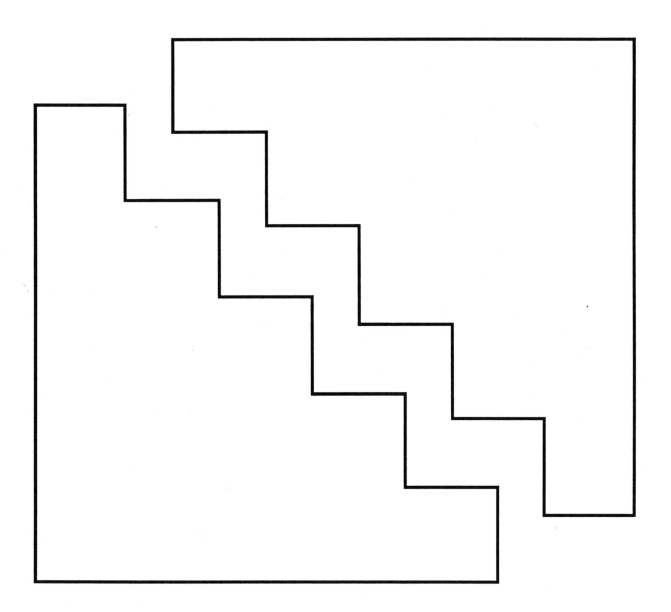

More For You
• Work with a group.
• Find other ways to build stairs using 3 pieces.
• Trace around each way you find.
• Share with other groups.

Making Rectangles

Name _____

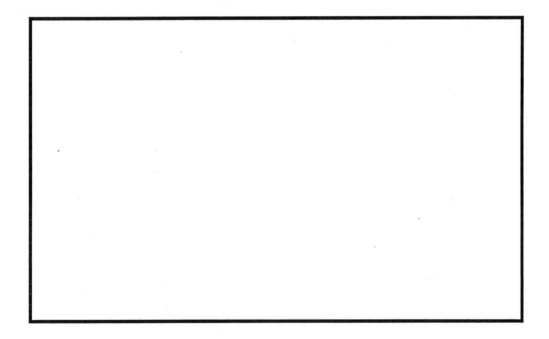

Use 3 pentomino pieces to make this rectangle.
Find as many ways as you can to make the rectangle.
Record each way you find on graph paper.

More For You

- Compare the ways you made the rectangle with the ways a friend did.
- Did you each make the rectangle in the same way?

Problem Solving with Pentominoes
© 1992 Learning Resources, Inc.

Making Large Rectangles

Name _____

Use 4 pentomino pieces to make this rectangle.
Find at least four ways you can to make the rectangle.
Draw a picture of some of the ways you find below.

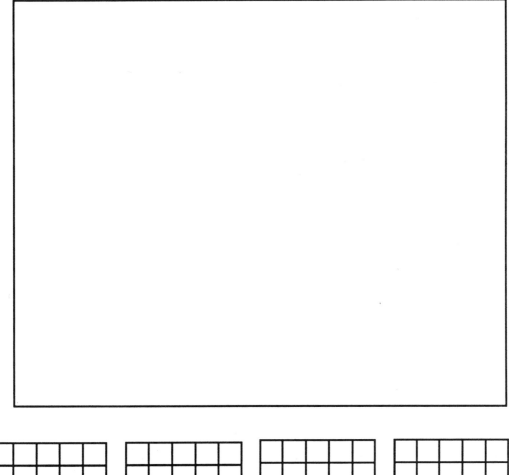

More For You
• Can you find more ways to make this rectangle?
• Use graph paper to record any other ways you find.

A Dog

Name _____

Cover the dog with 5 pentomino pieces.
Trace around the pieces to show the answer.

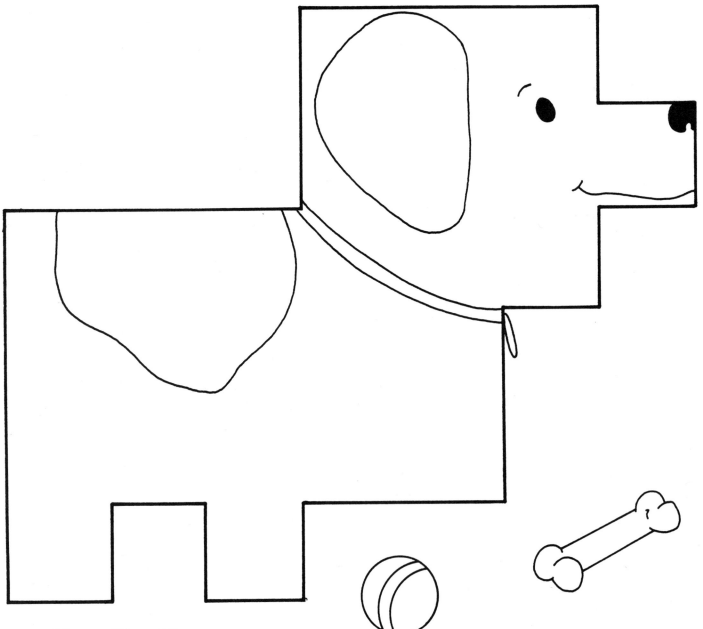

More For You
• Find other ways to cover the dog.
• Trace around the pieces you used on the back of this paper.

52

Problem Solving with Pentominoes
© 1992 Learning Resources, Inc.

A Butterfly

Name _____

Cover the butterfly with 6 pentomino pieces.

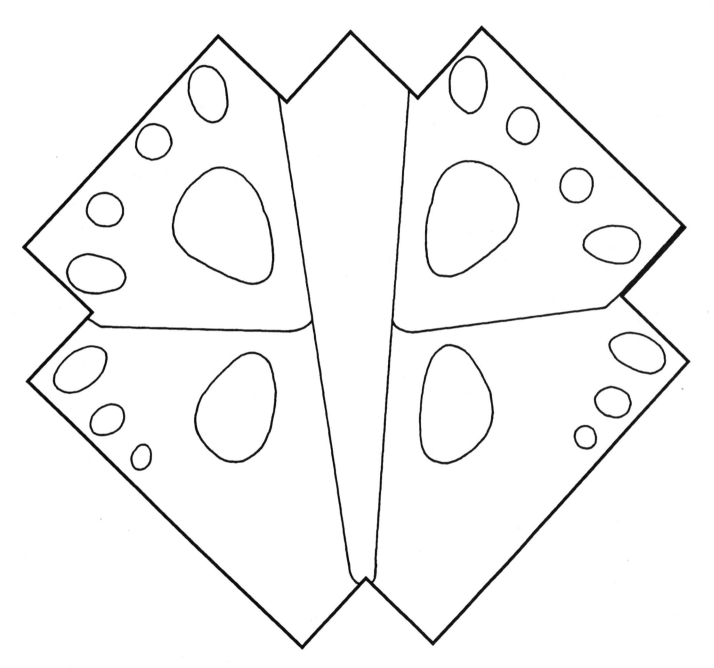

More For You
• Take two pieces off the puzzle.
• Put these with the other pentomino pieces.
• Ask a friend to find the missing pieces and cover the puzzle.

A Flag

Name _____

Cover the flag with 5 pentomino pieces.

More For You
• Make your flag another way. Your flag can have a funny shape.
• Trace around it and color it.

Problem Solving with Pentominoes
© 1992 Learning Resources, Inc.

Making Squares

Name _____

This square can be made in different ways using pentomino pieces. Find one way to make this square.

More For You
- Find other ways to make the square.
- Record each way you find on graph paper.

A Square Game

Name _____

Play with a partner.
In turn, players choose a pentomino piece to put on the game board.
Pieces cannot overlap or extend beyond the game board.
The last player to be able to put a piece on the game board wins.

More For You
•Can you find a strategy to help you win?

Problem Solving with Pentominoes
© 1992 Learning Resources, Inc.

Share at Home

Name _____

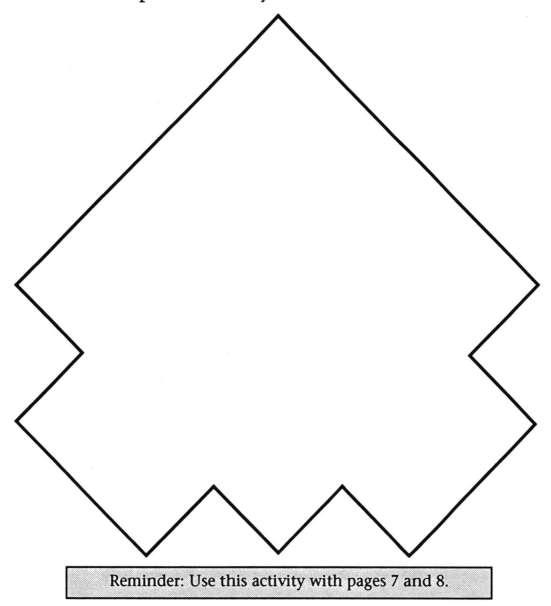

Dear Family,

We have been using pentomino pieces in school to cover shapes. Have your child cut out the pentomino pieces on pages 7 and 8. Then ask the child to use some pieces to cover the shape below.

Cover the shape with pentomino pieces.
Trace around each piece to show your answer.

Reminder: Use this activity with pages 7 and 8.

Symmetry

Teacher's Notes

In this section, children will explore symmetry with their pentomino pieces. Children will work to determine which pieces have line symmetry and which have rotational symmetry.

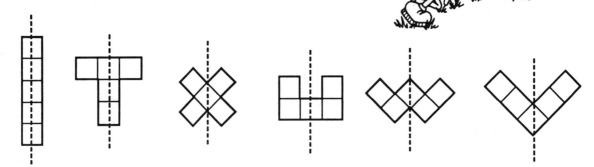

These pieces have *line* symmetry.

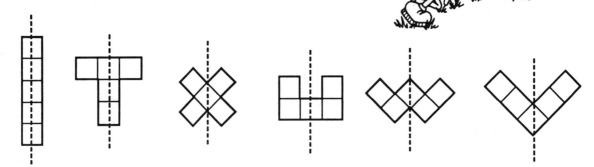

These pieces look the same when they are rotated a half turn (180°). We say these pieces have *turn* or *rotational* symmetry.

The remaining pentominoes are unsymmetrical.

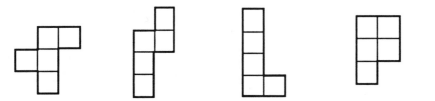

Knowing which pieces are symmetrical is valuable information for solving pentomino puzzles. When covering a shape, the symmetrical pieces are easier to place as they do not need to be turned around to fit. Pieces without line or rotational symmetry could fit on a puzzle in eight different ways. For example:

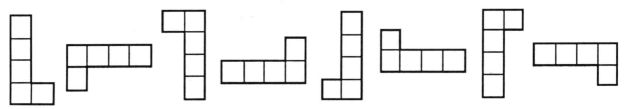

Problem Solving with Pentominoes
© 1992 Learning Resources, Inc.

When solving pentomino puzzles, place the symmetrical pentominoes on the puzzle first since there are fewer ways in which they could fit. Place the unsymmetrical pieces last.

Warm-Up

Provide each child with a sheet of 8 1/2" x 11" paper. Ask children to fold the paper in half vertically and compare the two parts. (The parts are the same or they match.) Then have children fold the paper in half horizontally and compare again. (These parts match as well.) Let children use rulers to draw the lines of symmetry on their paper. Then pose this problem: **Are there any other ways you could fold the paper to find two matching parts?** Children may experiment to find that folding on a diagonal does not produce symmetrical halves. Conclude with children that the paper has two lines of symmetry.

 ## Using the Pages

Pages 61 & 62

Symmetry of Pentomino Pieces: Give each child copies of pages 7, 8, 61, and 62. Have children cut out the pentomino pieces on pages 7 and 8. Help children fold each pentomino cutout to find if it has symmetry. Children can draw any lines of symmetry they find on their cutouts, then record the information on pages 61 and 62. Give children large sheets of construction paper on which to paste the pentomino cutouts. You may also want to give children a sheet of paper which they can use to explore the symmetry of a rectangle in the *More For You* section.

Note: It is recommended that pages 61 and 62 be completed together in one class session. However, if the age and attention span of the children makes this difficult, complete these pages in two sessions. Save the construction paper activity to use in the Wrap-Up lesson.

Page 63

Symmetry of a Shape: Give children scissors and another sheet of paper. Ask them to cover the shape with pentomino pieces, trace around the shape on their extra sheet of paper, and cut out the shape. Ask children to predict whether the shape is symmetrical before they do any folding. After children record their responses, discuss their findings.

Page 64

Turn Symmetry: Trace around the pentomino pieces **F** and **I** on a sheet of acetate and display them on the overhead projector. Ask volunteers to put the **F** and **I** on the outlines, then rotate the pieces a half turn (180°). Ask **Which piece fits back in its outline?** (I) Explain that this piece has *turn* or *rotational symmetry*. It will fit back in its outline after a half turn. Give pairs of children a set of pentominoes, copies of page 64, and sheets of paper. Working together, have children trace each piece, check for rotational symmetry, and record their responses on their activity page. In the *More for You* section, children will find which pieces have turn symmetry after a quarter turn. If necessary, review how to make a quarter turn.

Page 65

More Turn Symmetry: Read the directions with children and ask them to predict whether the shape has rotational symmetry before they rotate their pentomino pieces. Children can also test for rotational symmetry by tracing around the shape, cutting it out, and folding it vertically into two parts. After cutting along the fold line, tell children to rotate the pieces to see whether they match. Point out that matching pieces indicate that the shape has rotational symmetry.

Page 66

Share at Home: Give each child a copy of pages 7, 8, and 66. Read the instructions and tell children to use their cutouts to complete the activity at home with family members.

Wrap-Up

Give pairs of children a set of pentominoes and the construction paper on which they pasted their cutouts of symmetrical pieces. Ask children to sort the pieces to show which have no lines of symmetry, and which have one or more lines of symmetry. Then ask children to identify which pieces with line symmetry also have rotational symmetry. Discuss the findings with the children.

Next, form the shape below using the **F**, **T**, **P**, and **X,** and trace around the shape on a sheet of acetate. Display the shape on the overhead projector along with the corresponding pentomino pieces. Ask volunteers to tell which pieces they would place on the puzzle first, second, third, and fourth. Listen to children's descriptions to see whether they understand how symmetrical pieces are more easily placed within a puzzle.

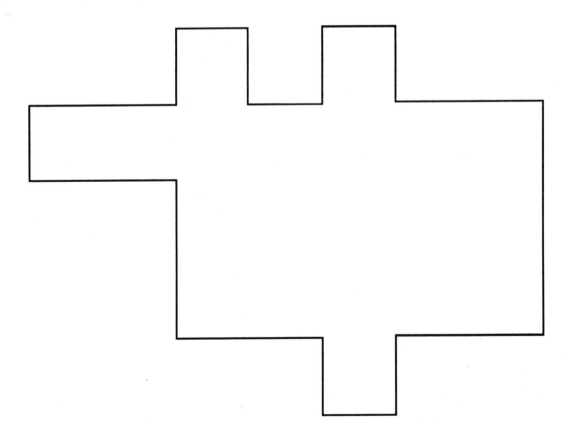

Symmetry of Pentomino Pieces 1

Name _____

Cut out each pentomino piece on page 7.
Find each piece and fold it.
Is it symmetrical? Circle YES or NO.
If it is symmetrical, tell how many lines are symmetrical.

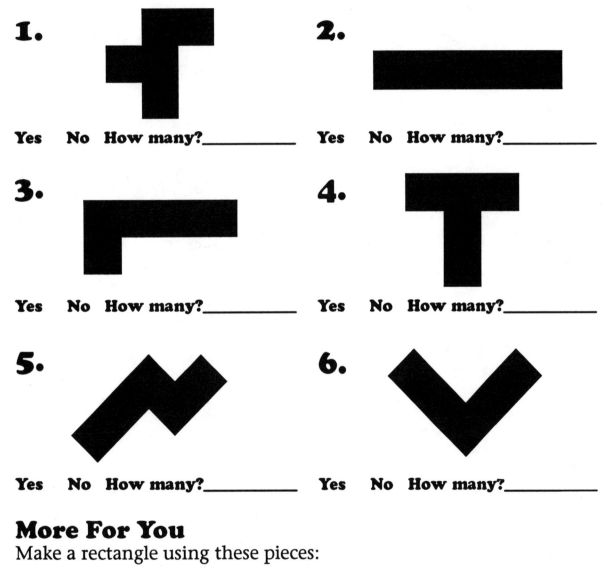

1.

Yes   No   How many?_____

2.

Yes   No   How many?_____

3.

Yes   No   How many?_____

4.

Yes   No   How many?_____

5.

Yes   No   How many?_____

6.

Yes   No   How many?_____

More For You

Make a rectangle using these pieces:

- Trace around and cut out the rectangle.
- Does it have a line or lines of symmetry? YES NO
- How many? _____

Symmetry of Pentomino Pieces 2

Name _____

Cut out each pentomino piece on page 8.
Find each piece and fold it.
Is it symmetrical? Circle YES or NO.
If it is symmetrical, tell how many lines are symmetrical.

1.

Yes   No   How many?_____

2.

Yes   No   How many?_____

3.

Yes   No   How many?_____

4.

Yes   No   How many?_____

5.

Yes   No   How many?_____

6.

Yes   No   How many?_____

More For You

• Make a square using these pieces:

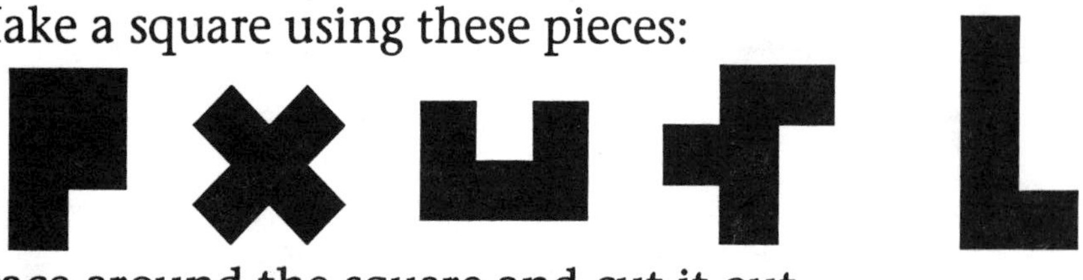

• Trace around the square and cut it out.
• Does it have a line or lines of symmetry? YES NO
• How many? _____

Problem Solving with Pentominoes
© 1992 Learning Resources, Inc.

Symmetry of a Shape

Name _____

Cover the shape with pentomino pieces.
Trace around the shape on another piece of paper.
Cut it out and fold it. Is it symmetrical? Circle YES or NO
How many lines of symmetry does it have? _____

More For You
• Make a symmetrical shape with your pentomino pieces.
• Trace, cut, and fold the shape to test it.

Problem Solving with Pentominoes
© 1992 Learning Resources, Inc.

Turn Symmetry

Name _____

Some pentomino pieces will fit back in their outline after a half turn.
These pieces have *turn symmetry*.
Trace around each pentomino piece on a sheet of paper.
Turn it a half turn.
Does it fit back in its outline? Check YES or NO.

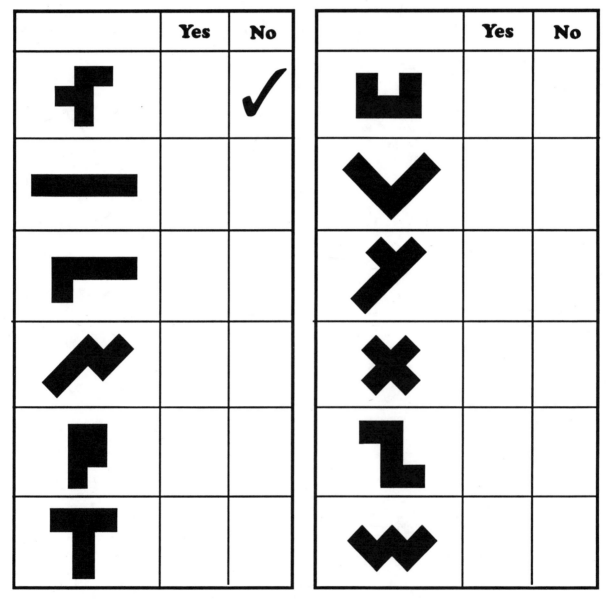

	Yes	No
		✔

	Yes	No

More For You

- Look at the pieces that have turn symmetry.
- Which pieces fit back in their outline after a quarter turn?
- Try each piece.

More Turn Symmetry

Name _____

Make this shape with your pentomino pieces.
Turn it a half turn.
Does it fit back in its outline? Circle YES or NO.

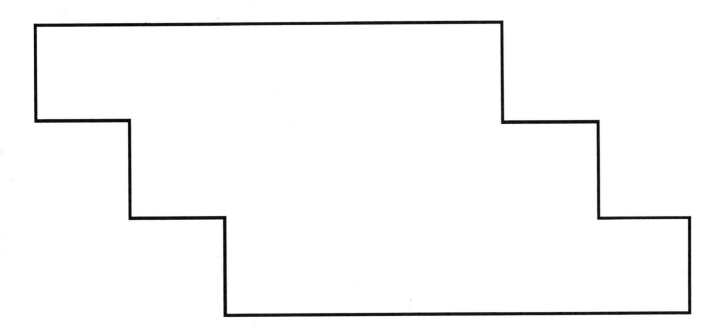

More For You

• Build this shape:

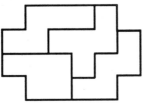

• Trace around it on another sheet of paper.
• Turn the shape a half turn.
• Does it fit back in its outline?

Share at Home

Name _____

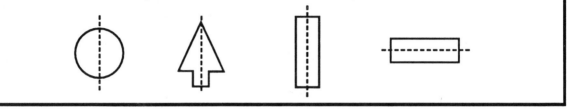

Dear Family,

We have been using pentomino pieces in school to learn about symmetry. Have your child cut out the pentomino pieces on pages 7 and 8. Make the shapes and then find the lines of symmetry. A figure has symmetry if it can be folded so that the two parts match exactly. These figures have symmetry:

Make each shape with pentomino pieces.
Trace around it on another sheet of paper and cut it out.
Is it symmetrical? Circle YES or NO.
How many lines of symmetry does it have?

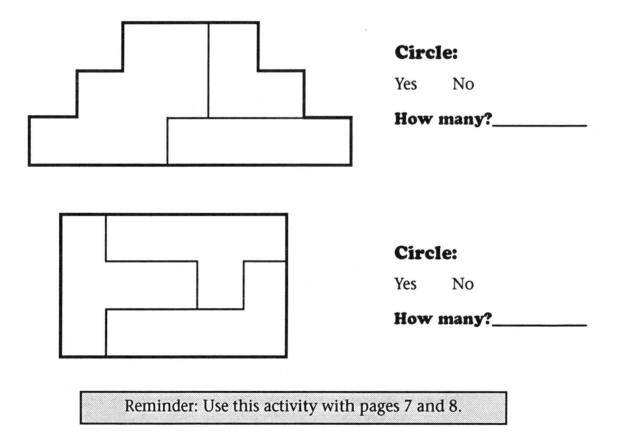

Circle:

Yes   No

How many?_____

Circle:

Yes   No

How many?_____

Reminder: Use this activity with pages 7 and 8.

Problem Solving with Pentominoes
© 1992 Learning Resources, Inc.

Perimeter & Area

Teacher's Notes

Premier Pentominoes are scored to make them especially suited for finding the perimeter and area of a shape. Each piece is made of five one-inch squares. The pieces can be used with the inch graph paper (page 93).

Perimeter and area of a shape can be easily found using the scored pentomino pieces. Children can count the units along each side to find perimeter, and the square units within each shape to find area. If perimeter and area are new to the children, permit some free exploration time with inch squares commercially prepared or cut from tagboard. Let children cover a 3"x 5" card with the inch squares. Tell them to place the squares on the card so there is no overlap and no gaps. Ask for an estimate of the perimeter. Then have children count the sections along the sides to find the perimeter. Next, children can estimate and count to find the area. Encourage them to estimate the perimeter and area of each shape in this section before completing each activity.

Warm-Up

Prepare inch graph paper on a sheet of acetate. Display the acetate on the overhead projector, align a pentomino piece on the graph paper, and trace around it as shown on the right. Ask children to estimate how many units make up the outside of the shape.

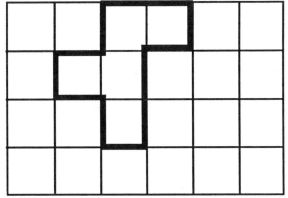

Demonstrate by marking your starting spot and counting around the perimeter. Similarly, ask children to estimate the number of square units in the shape, then count the units together. Discuss the terms *perimeter* and *area* if these are new to the children.

Using the Pages

Page 69

Perimeter of Pentomino Pieces: Demonstrate how the perimeter of a shape is found (see Warm-Up). Then give pairs of children a set of pentominoes and copies of page 69. (Both the *Premier Pentominoes* and pages 7 and 8 are specially scored so children can count to find perimeter.) It may be helpful for children to put a mark on the section where they begin their count. In the *More For You* section, children will identify the pentomino **P** as having the smallest perimeter. Unlike the other pieces, four of the squares in the **P** share two common sides. This more compact piece has a smaller perimeter.

Page 70
Perimeter of Shapes: Provide each child or each pair of children with a set of pentominoes to find the perimeter of these shapes. Once children cover each shape with pentominoes, they can count the units to determine the perimeter.

Page 71
Perimeter of a Rectangle: Read the directions and let children work independently. Point out that this 4"x 5" unit rectangle can be covered with 25 different sets of pentomino pieces. Help children realize that since the size of the rectangle does not change with each new set of pieces, the perimeter remains the same.

Page 72
Make a Figure: Through experimentation, children can construct a figure with a given perimeter. Ask children to first estimate how many pieces they would use for the figure, then arrange the pieces on the graph section of the activity page. To help children revise their figure to come closer to the given perimeter, ask questions such as: **Is the perimeter of your figure close to 20? Do you need to add another piece to your figure, or remove a piece to come closer to a perimeter of 20? How could you adjust your pieces to get a perimeter of 20?**

Page 73
Area: Children can cover the shapes with their pentomino pieces, then count the squares to find the area of each shape. Demonstrate how the area of a shape can be found by counting square units. Then have children complete the exercises on the page.

Page 74
Area of a Square: As with the perimeter of a rectangle on page 71, children should realize that the area of a square does not change even when the square is covered in different ways. The area remains the same. When children are ready to complete the *More for You* section, ask, **How many squares are in each pentomino piece? (5) How can you use that information to help you find the area of a shape with two pentomino pieces, or three pentomino pieces?** (Count by fives.)

Page 75
Share at Home: Give each child copies of pages 7, 8, and 75. Read the instructions and tell children to use their cutout pentomino pieces to complete the activity at home with a family member.

Wrap-Up
To review perimeter and area, trace around the outside of several shapes made from two or more pentomonio pieces on sheets of acetate. Display the shapes one at a time on an overhead projector. Ask volunteers to come to the overhead projector, cover the shape with pentomino pieces, and count to find the perimeter and area of the shape.

Perimeter of Pentomino Pieces

Name_____

The number of units around a shape is its *perimeter*.
This pentomino piece has a perimeter of 12 units.
Find the perimeter of each pentomino piece.
Write the perimeter in the chart.

= 1 unit

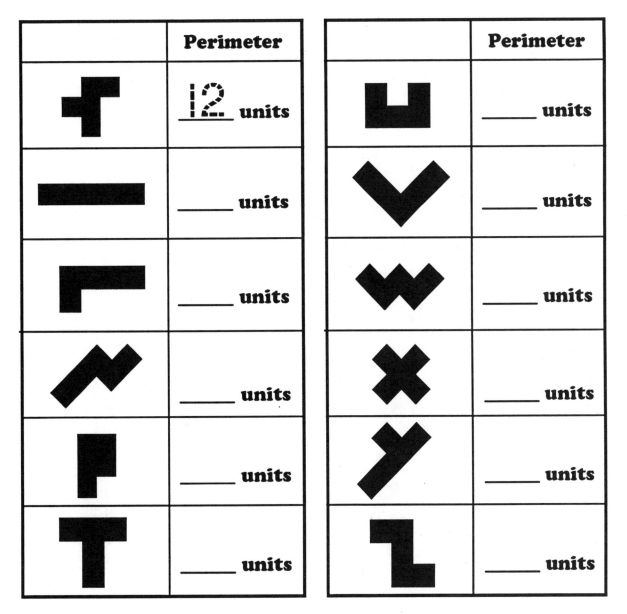

	Perimeter		Perimeter
	_12_ units		_____ units
	_____ units		_____ units
	_____ units		_____ units
	_____ units		_____ units
	_____ units		_____ units
	_____ units		_____ units

More For You
• Which pentomino piece has the smallest perimeter?
• Is this piece different from the others? Share what you think.

Perimeter of Shapes

Name _____

Cover each shape with pentomino pieces.
Give the perimeter of each shape.

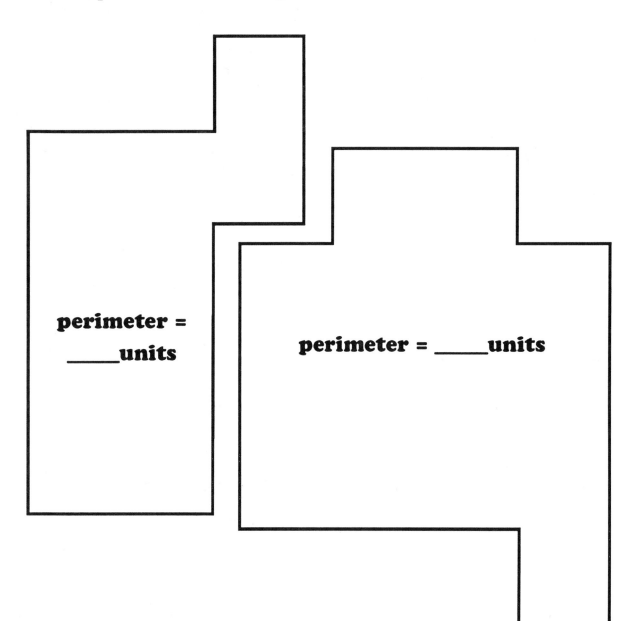

perimeter =
_____units

perimeter = _____units

More For You
• Use any 2 pentomino pieces to make a shape with the smallest possible perimeter.
• Trace around your shape on a sheet of paper.

70

Problem Solving with Pentominoes
© 1992 Learning Resources, Inc.

Perimeter of a Rectangle

Name _____

Cover the rectangle with pentomino pieces.
Give the perimeter. P = _____ units.
Cover the rectangle with another set of pentomino pieces.
Give the perimeter. P = _____ units.

More For You
- If you cover the rectangle with another set of pentomino pieces, what will the perimeter be?
- Try it.

Make a Figure

Name _____

Use any of your pentomino pieces.
Make a shape with a perimeter of 20.
Trace around your shape below.

More For You
• Use these pentomino pieces:

• Make a shape with a perimeter of 22.
• Trace around your shape on another sheet of paper.
• Is your shape the same as a friend's shape?

Problem Solving with Pentominoes
© 1992 Learning Resources, Inc.

Area of Shapes

Name _____

The number of square units inside a shape is its area.
This pentomino piece has an area of 5 square units.
Cover each shape below with pentomino pieces.
Give the area of each shape.

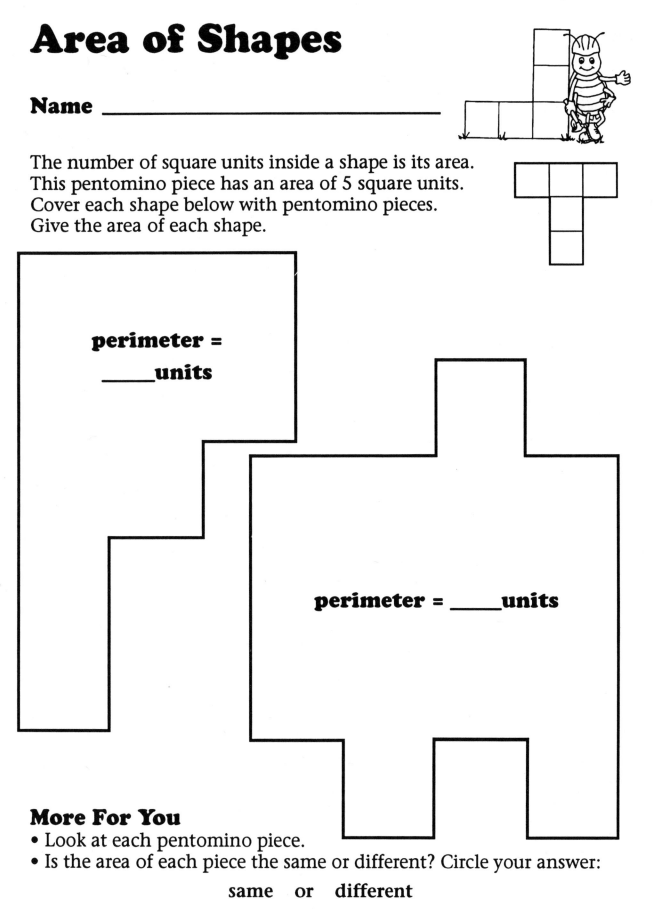

**perimeter =
_____units**

perimeter = _____units

More For You
- Look at each pentomino piece.
- Is the area of each piece the same or different? Circle your answer:

same   or   different

- Count to check.

Area of a Square

Name _____

Cover the square with pentomino pieces. Give the area.
If you cover the square another way, will the area remain the same?
Guess, then check.

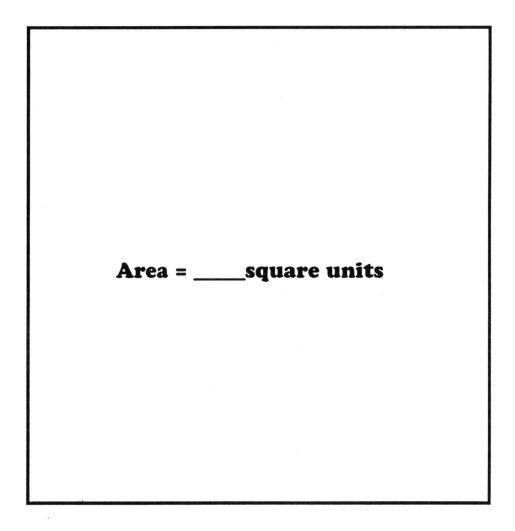

Area = _____square units

More For You
What will the area be for these shapes:

- A 2 pentomino shape?   _____ units.

- A 3 pentomino shape?   _____ units.

- A 12 pentomino shape?   _____ units.

74

Share at Home

Name _____

Dear Family,

We have been using pentomino pieces in school to learn about perimeter and area. Have your child cut out the pentomino pieces on pages 7 and 8. Then help your child use the pentomino pieces to find the area and perimeter of the shapes below.

☐ = 1 square unit   ☐ → ┆ = 1 unit

Cover each shape with pentomino pieces.
Give the perimeter and area of each shape.

1.

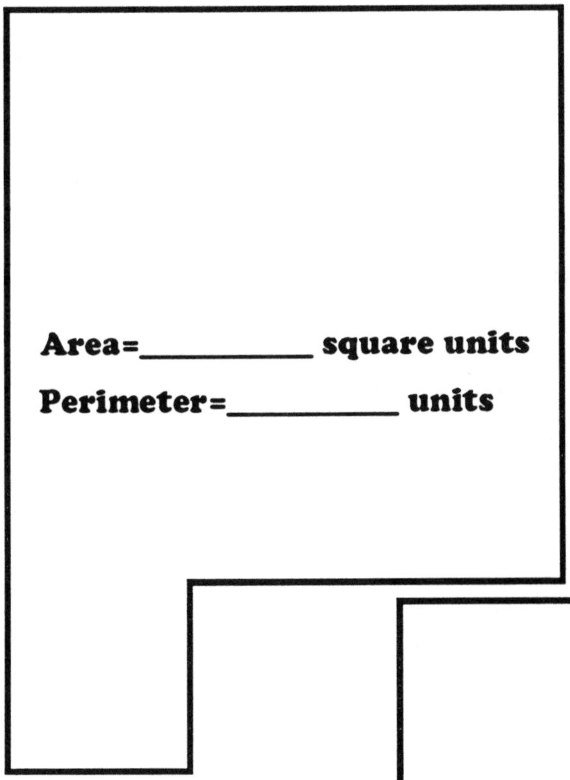

Area=_____ square units

Perimeter=_____ units

2.

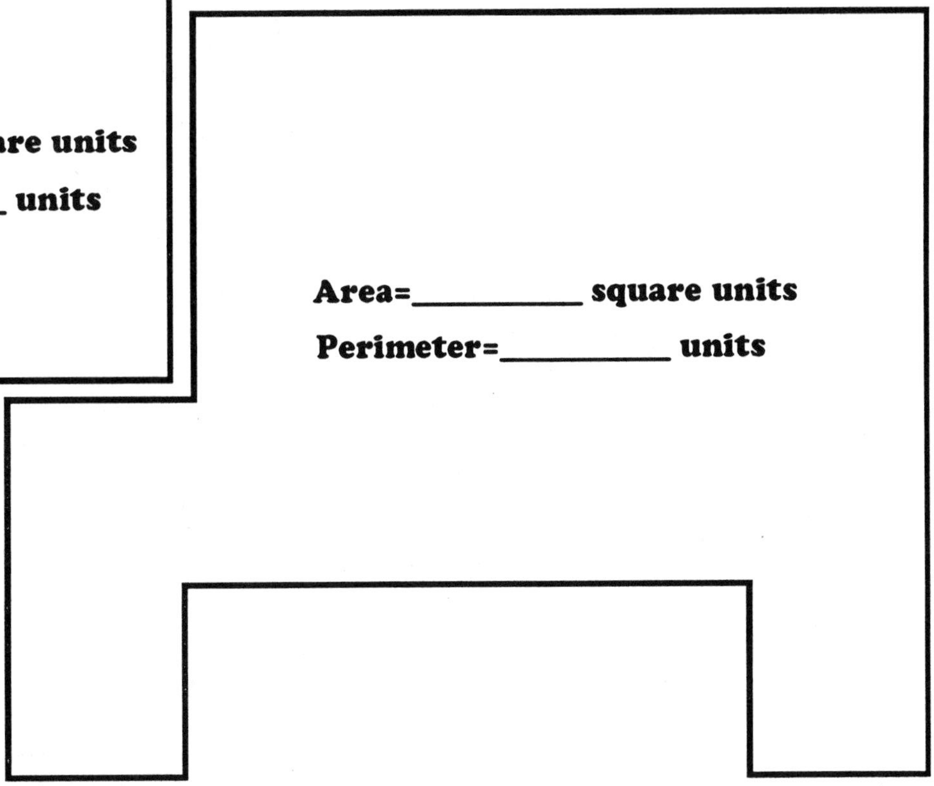

Area=_____ square units

Perimeter=_____ units

Reminder: Use this activity with pages 7 and 8.

Congruence & Similarity

Teacher's Notes

In this section, children will use pentomino pieces to explore congruence and similarity. *Congruent* shapes have the same shape and size. *Similar* shapes have the same shape but are proportional in size. In similar shapes, corresponding angles are congruent.

Congruent Shapes   Similar Shapes

Children will usually understand the concept of congruence before similarity. Using pentomino pieces, children can cover congruent shapes, then place one set of pieces on top of the other to help them visualize the congruence of the shapes. To understand similarity, children have to think in terms of enlarging or reducing a shape.

Warm-Up

In this activity, children will learn how to recognize congruent and similar figures. Form these two *congruent* shapes on an overhead projector.

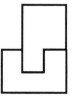

Ask a volunteer to place one set of shapes on top of the other and compare their size and shape (same). Tell children that when shapes have the same size and shape, they are congruent. Now form these two shapes on an overhead projector.

Children need to recognize that these are the same shape but not the same size. Help a volunteer place the smaller shape on top of the larger shape to compare the angles. By lining up the longest side of both shapes, children can compare corresponding angles of both shapes and determine whether they match or are congruent. In this section, children will compare the angles of similar shapes along with the lengths of the sides to determine similarity.

Using the Pages

Page 79

Congruent Shapes: Read the directions with children and let them work independently. Help them prove congruence in this activity by covering the two shapes with pentomino pieces, then placing one set of pieces on top of the other for a visual comparison. Help children discover that even when congruent shapes are covered with different sets of pieces, the shapes remain congruent. By layering the sets of pieces, children can usually determine congruence.

Page 80

Making Congruent Shapes: Do the first problem together with the children to be sure they understand how to complete the activity. Children are to use the two pieces pictured to form a shape that is congruent to either Shape A or Shape B. Let children work independently or in small groups.

Page 81

Similar Shapes: Direct children to cover each shape with pentomino pieces, then count to find the perimeter of each shape. **What do you notice about the perimeter of each shape?** (The perimeter of the large shape is two times the perimeter of the small shape.) Help children compare the corresponding sides of each shape to find that they are doubled in size. Ask children to place the small shape on top of the large shape and match each of the corresponding angles. **What do you notice about the corresponding angles of each shape?** (The corresponding angles are the same; they are congruent.)

Page 82

Doubling Pentominoes: Provide pairs of children with copies of page 82, inch graph paper (page 93), and a set of pentomino pieces. It might be easier for children to tape together two sheets of graph paper on which they can record a pentomino and a doubled pentomino. (Each pentomino shape can be doubled except for **V** and **X**.)

You may want to draw the doubled pentomino on graph paper together with the children. To do so, prepare inch graph paper on a sheet of acetate and display it on an overhead projector. First, have children trace around the pentomino piece on their graph paper. Then, as they follow along, draw the doubled pentomino on an overhead projector using the illustration on page 82 as a guide. As children cover each shape and find the perimeter for each shape, they will notice that the perimeter of the larger shape is twice the perimeter of the smaller shape.

Page 83

Tripling Pentominoes: This page uses the same format as page 82. On this page, the children will draw a single pentomino and a tripled pentomino on graph paper and cover them with pentomino pieces. After finding the perimeter of each shape, children will notice that the perimeter of the large shape is three times the perimeter of the small shape. Children can use this information to predict the perimeter of other tripled pentominoes.

Page 84

Share at Home: Give each child a copies of pages 7, 8, and 84. Read the instructions and tell children to use their cutout pentomino pieces to complete the activity at home with a family member.

Wrap-Up

Give groups of three children several sheets of inch graph paper (page 93), two sheets of oaktag, scissors, paste, and one set of pentominoes. Help children trace the four figures below on the graph paper. Children can cut out the figures and paste those that are congruent on one sheet of oaktag, as shown.

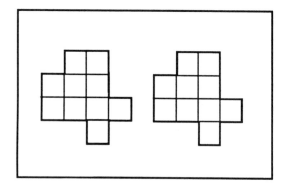 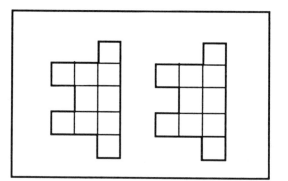

Children can use one set of pentominoes to cover each shape. Ask children to compare the two shapes on each tagboard and prove why they are either congruent or similar. Let each group present its arguments to the whole class.

Congruent Shapes

Name _____

Two shapes are congruent when they are the same shape and size.

Cover each shape with pentomino pieces.
Put one set of pieces on top of the other to match.
Are these shapes congruent? Circle your answer: YES or NO

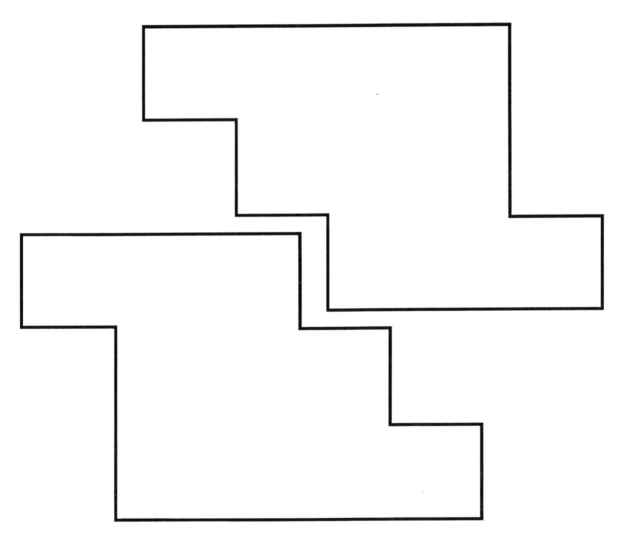

More For You
- Make another pair of congruent shapes using your pentomino pieces.
- How can you prove these shapes are congruent?

Making Congruent Shapes

Name _____

Compare shapes A and B with the shapes you will form in each problem.

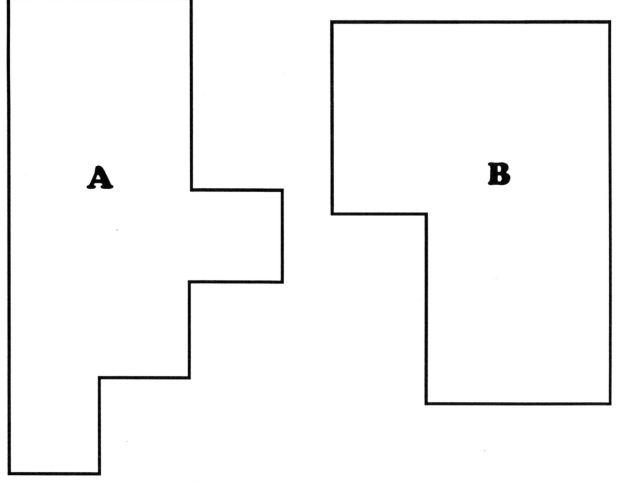

Use each set of pieces to make one of the shapes shown.
Write in the space whether your shape is congruent to A or B.

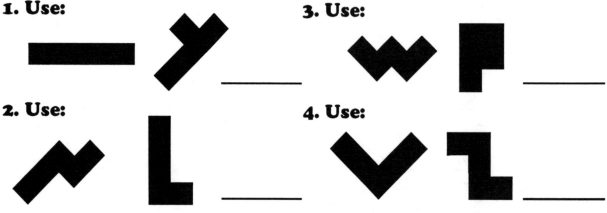

1. Use: _____

2. Use: _____

3. Use: _____

4. Use: _____

Problem Solving with Pentominoes
© 1992 Learning Resources, Inc.

Similar Shapes

Name _____

These two shapes are similar.
They have the same shape.
The large one is twice the size of the small one.
Cover each shape with pentomino pieces.
Find the perimeter of each shape.

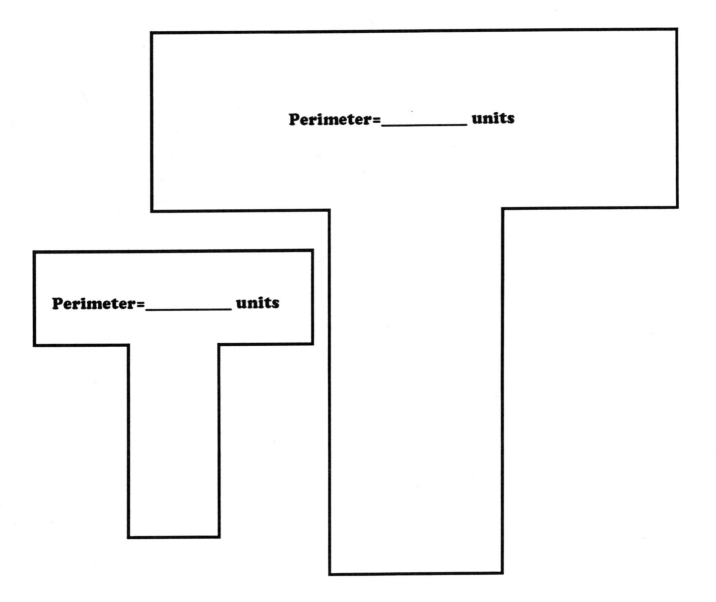

Perimeter=_____ units

Perimeter=_____ units

More For You

• How can you tell these two shapes are similar?

Doubling Pentominoes

Name _____

Each pair of pentomino shapes is similar.
The large one is twice the size of the small one.
Copy each shape on inch graph paper (page 93).
Cover each shape with pentomino pieces.
Find the perimeter of each shape.

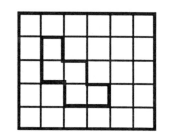

Perimeter=_____ units

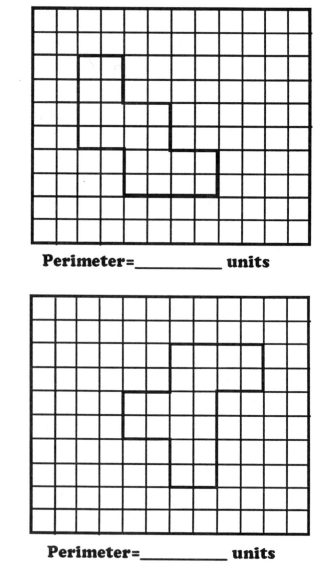

Perimeter=_____ units

Perimeter=_____ units

Perimeter=_____ units

More For You

- Look at the perimeters of each pair of similar shapes. What do you notice?
- Can you predict the perimeter of other pentominoes that are doubled?

Problem Solving with Pentominoes
© 1992 Learning Resources, Inc.

Tripling Pentominoes

Name _____

This pair of pentomino shapes is similar.
The large one is three times the size of the small one.
Tape 2 sheets of inch graph paper (page 93) together.
Copy each shape below on the graph paper.
Cover each shape with pentomino pieces.
Find the perimeter of each shape.

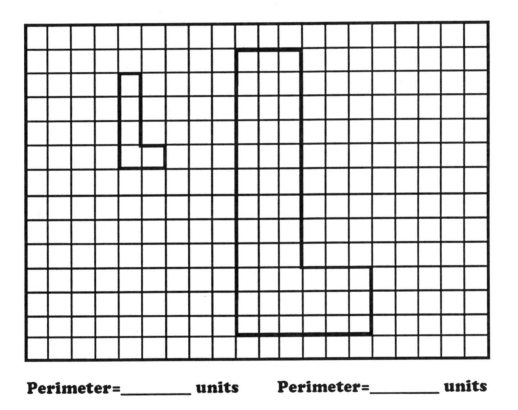

Perimeter=_____ units   **Perimeter=_____ units**

More For You

• Look at the perimeter you recorded above.
• Can this help you predict the perimeter of other tripled
  pentomino shapes?

Share at Home

Name _____

Cover the shape with two layers of pentomino pieces.
Are the layers congruent?

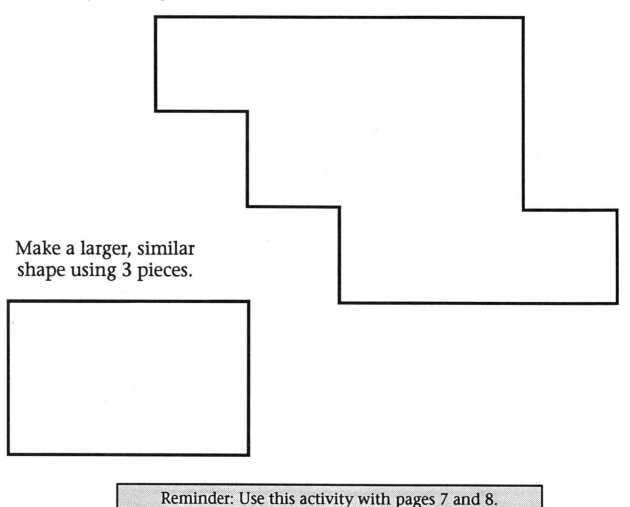

Make a larger, similar shape using 3 pieces.

Reminder: Use this activity with pages 7 and 8.

Pentomino Challenges

Teacher's Notes

Children will use their problem-solving skills with the challenging puzzles presented in this section. Let children work in pairs or small groups to complete the activities.

F L P U

Warm-Up

Display the shape above on an overhead projector along with the pentomino pieces listed. Ask children how they would begin to use the pieces to cover the shape. Give children time to complete the puzzle. Then let a volunteer show the solution on the overhead projector.

 **Using the Pages**

Pages 86-89

Use a Small Model, Animal Shapes, Making a Box, Share at Home: The first reduced puzzles children will build include all the interior lines. Children will need to position the pieces within the shape without the aid of the outer line. This is the first time children form a shape using all 12 pieces. For the box activity, demonstrate for children how to fold the **F** piece along each line, then form it into an open box. The **I** does not form a box, so it might be helpful to fold this piece together with the children. Ask children to point out the difference between the two boxes. (One forms a box with four sides and a bottom; the other forms a shape with five sides.)

Wrap-Up

Children can create their own reduced puzzles for classmates to solve. Provide children with quarter-inch graph paper (page 94) on which they can draw their puzzles. After the puzzles have been correctly drawn and shaded, ask children to cut them out and paste them on a 5"x 8" card. Keep these puzzle cards in the Math Corner for class members to solve.

Use a Small Model

Name _____

Use your pentomino pieces.
Make each small shape on a large sheet of paper.
Trace around the outside of each shape.

Use 10 pieces to make these buildings.

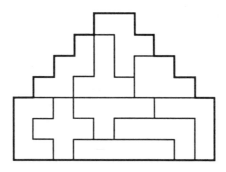
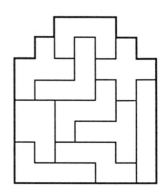

Use 12 pieces to make
these shapes.

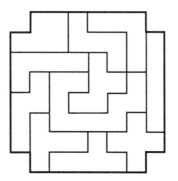
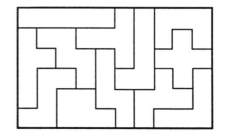

More For You
- Some pentomino shapes are formed with holes inside. These are holey shapes.
- Use your pieces to make this holey square.
- Trace around the square on page 93.

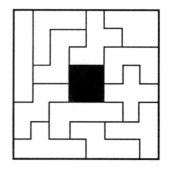

Problem Solving with Pentominoes
© 1992 Learning Resources, Inc.

Animal Shapes

Use your pentomino pieces.
Make the shapes using the pieces shown.
Trace around each shape on another sheet of paper.

Use:

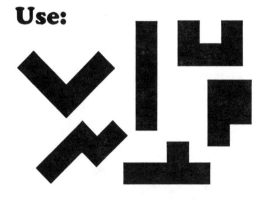
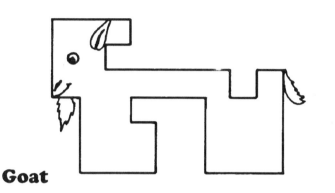

Goat

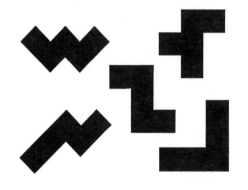
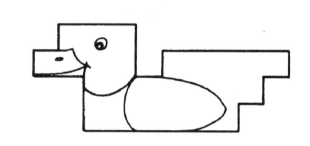

Duck

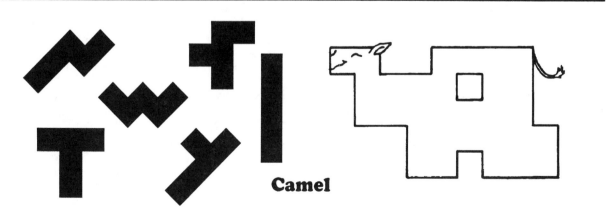

Camel

More For You

- Use some or all of your pentomino pieces to make an animal shape.
- Trace around your shape on a sheet of paper.

Making a Box

Name _____

Some of the pentomino pieces will form an open box when folded.
Cut out each pentomino piece on pages 7 and 8.
Fold each piece along the lines.
Complete the chart.

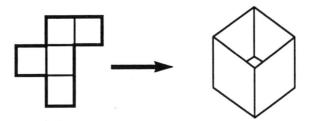

Open Boxes

	Yes	No		Yes	No
F	✓		U		
I			V		
L			W		
N			X		
P			Y		
T			Z		

More For You
• Look at the pieces that did not form an open box.
• Which of these pieces can be folded to form a box with four sides?

Problem Solving with Pentominoes
© 1992 Learning Resources, Inc.

Share at Home

Name _____

Use all 12 pentomino pieces to make each shape
The holes in a shape do not need to be covered

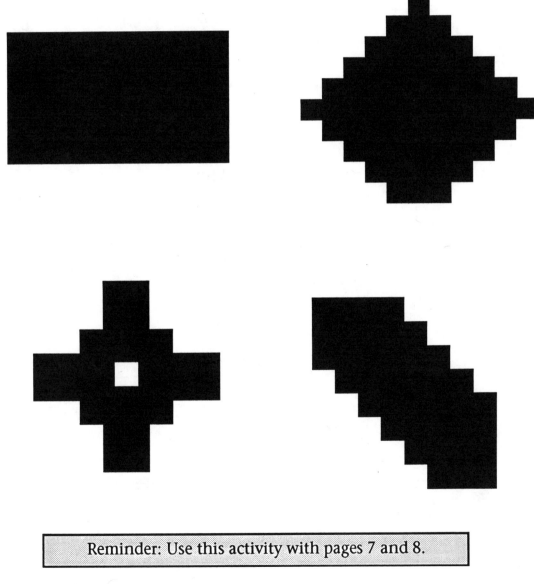

Reminder: Use this activity with pages 7 and 8.

Selected Solutions

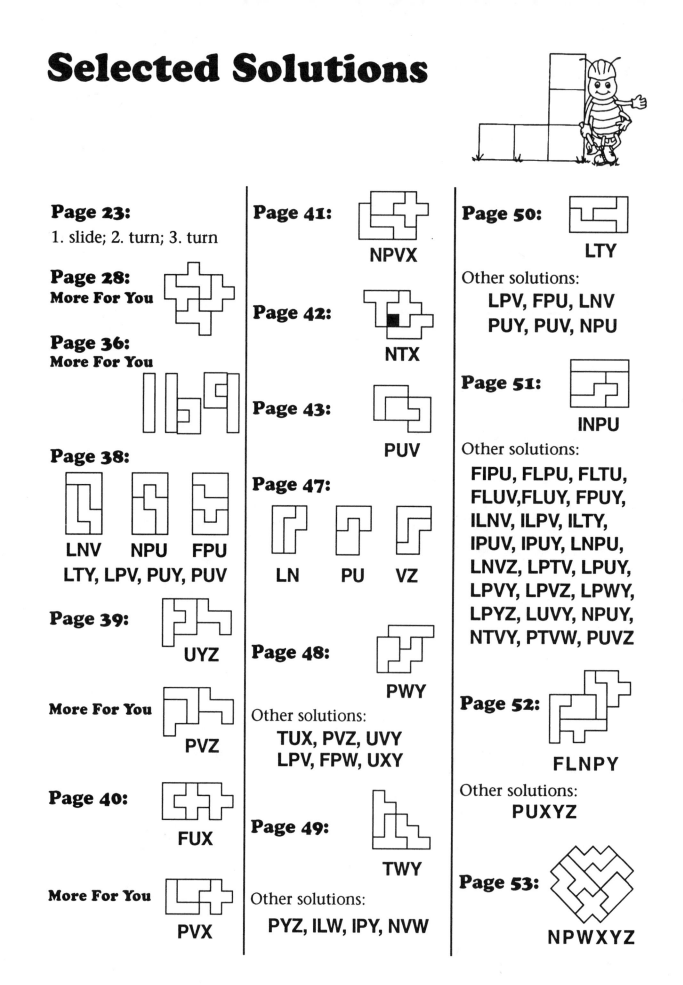

Page 23:
1. slide; 2. turn; 3. turn

Page 28:
More For You

Page 36:
More For You

Page 38:
LNV   NPU   FPU
LTY, LPV, PUY, PUV

Page 39:
UYZ

More For You
PVZ

Page 40:
FUX

More For You
PVX

Page 41:
NPVX

Page 42:
NTX

Page 43:
PUV

Page 47:
LN   PU   VZ

Page 48:
PWY

Other solutions:
TUX, PVZ, UVY
LPV, FPW, UXY

Page 49:
TWY

Other solutions:
PYZ, ILW, IPY, NVW

Page 50:
LTY

Other solutions:
LPV, FPU, LNV
PUY, PUV, NPU

Page 51:
INPU

Other solutions:
FIPU, FLPU, FLTU,
FLUV, FLUY, FPUY,
ILNV, ILPV, ILTY,
IPUV, IPUY, LNPU,
LNVZ, LPTV, LPUY,
LPVY, LPVZ, LPWY,
LPYZ, LUVY, NPUY,
NTVY, PTVW, PUVZ

Page 52:
FLNPY

Other solutions:
PUXYZ

Page 53:
NPWXYZ

Problem Solving with Pentominoes
© 1992 Learning Resources, Inc.

Selected Solutions

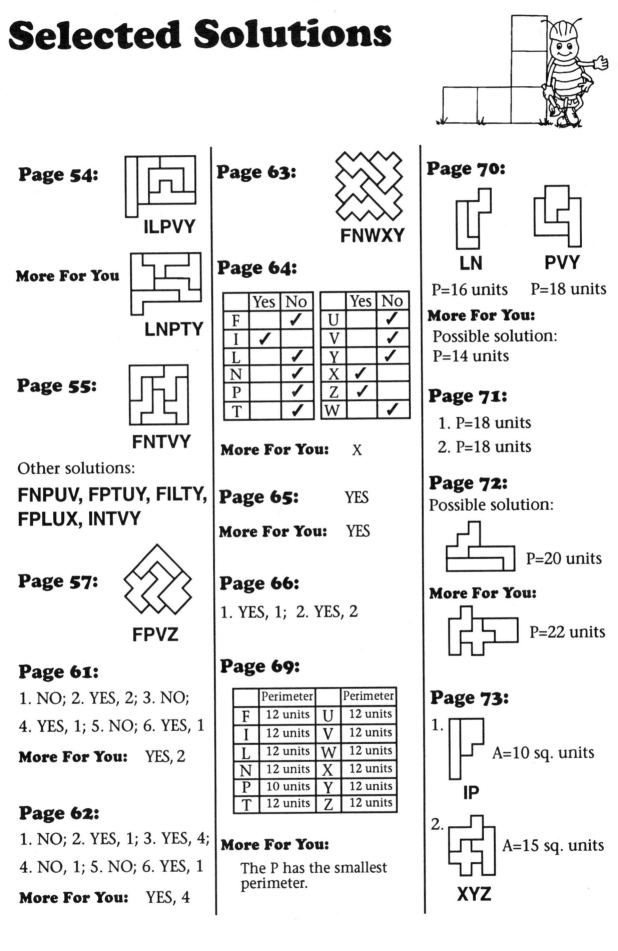

Page 54:

ILPVY

More For You

LNPTY

Page 55:

FNTVY

Other solutions:

FNPUV, FPTUY, FILTY, FPLUX, INTVY

Page 57:

FPVZ

Page 61:

1. NO; 2. YES, 2; 3. NO;

4. YES, 1; 5. NO; 6. YES, 1

More For You: YES, 2

Page 62:

1. NO; 2. YES, 1; 3. YES, 4;

4. NO, 1; 5. NO; 6. YES, 1

More For You: YES, 4

Page 63:

FNWXY

Page 64:

	Yes	No		Yes	No
F		✓	U		✓
I	✓		V		✓
L		✓	Y		✓
N		✓	X	✓	
P		✓	Z	✓	
T		✓	W		✓

More For You: X

Page 65:   YES

More For You: YES

Page 66:

1. YES, 1; 2. YES, 2

Page 69:

	Perimeter		Perimeter
F	12 units	U	12 units
I	12 units	V	12 units
L	12 units	W	12 units
N	12 units	X	12 units
P	10 units	Y	12 units
T	12 units	Z	12 units

More For You:

The P has the smallest perimeter.

Page 70:

LN   PVY

P=16 units   P=18 units

More For You:

Possible solution:
P=14 units

Page 71:

1. P=18 units

2. P=18 units

Page 72:

Possible solution:

P=20 units

More For You:

P=22 units

Page 73:

1.

A=10 sq. units

IP

2.

A=15 sq. units

XYZ

Selected Solutions

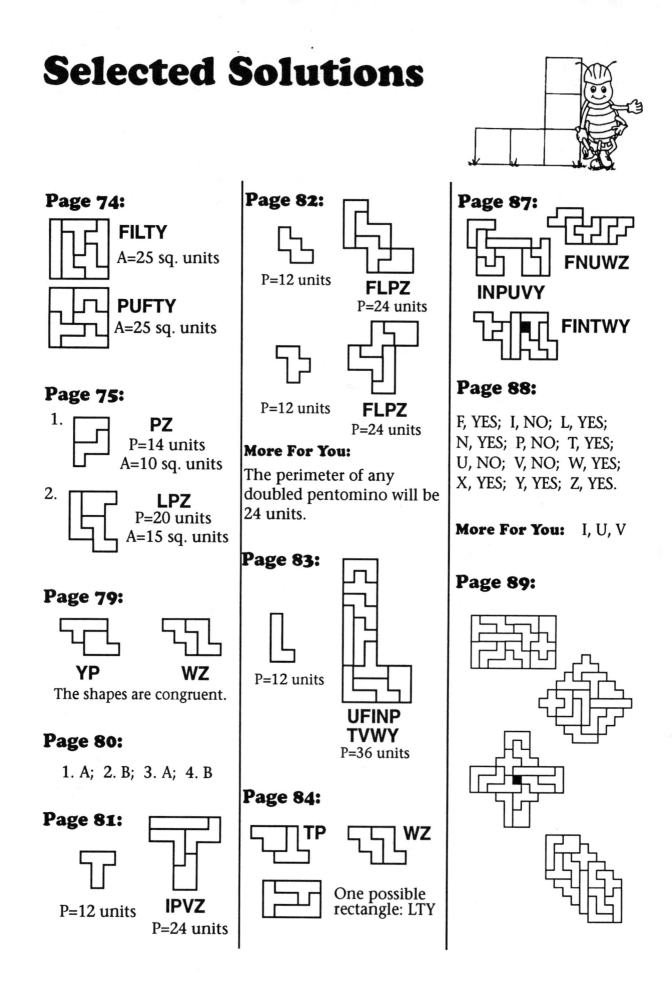

Page 74:

FILTY
A=25 sq. units

PUFTY
A=25 sq. units

Page 75:

1. **PZ**
P=14 units
A=10 sq. units

2. **LPZ**
P=20 units
A=15 sq. units

Page 79:

YP   **WZ**
The shapes are congruent.

Page 80:

1. A; 2. B; 3. A; 4. B

Page 81:

P=12 units

IPVZ
P=24 units

Page 82:

P=12 units

FLPZ
P=24 units

P=12 units

FLPZ
P=24 units

More For You:

The perimeter of any doubled pentomino will be 24 units.

Page 83:

P=12 units

UFINP TVWY
P=36 units

Page 84:

TP   **WZ**

One possible rectangle: LTY

Page 87:

FNUWZ

INPUVY

FINTWY

Page 88:

F, YES; I, NO; L, YES;
N, YES; P, NO; T, YES;
U, NO; V, NO; W, YES;
X, YES; Y, YES; Z, YES.

More For You: I, U, V

Page 89:

Problem Solving with Pentominoes
© 1992 Learning Resources, Inc.

Inch Graph Paper

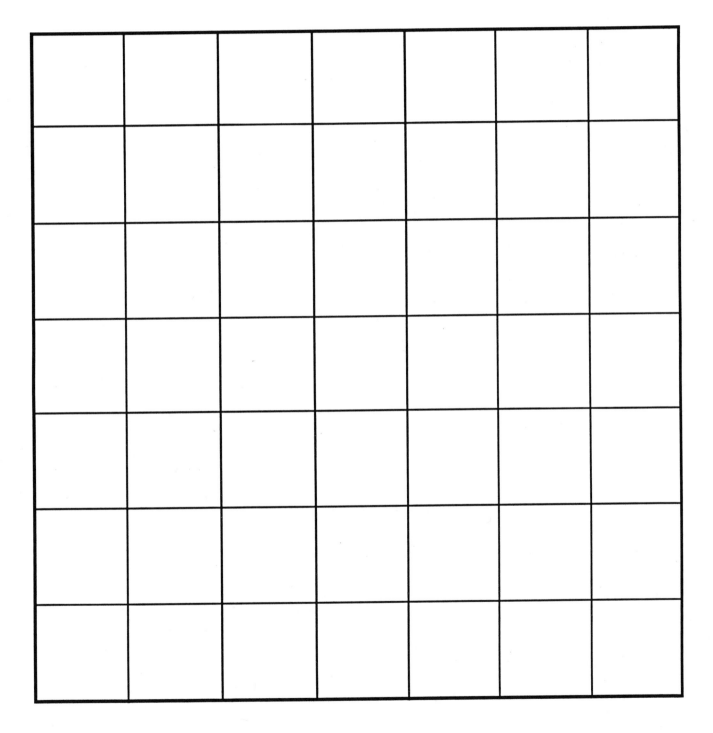

1/4 Inch Graph Paper

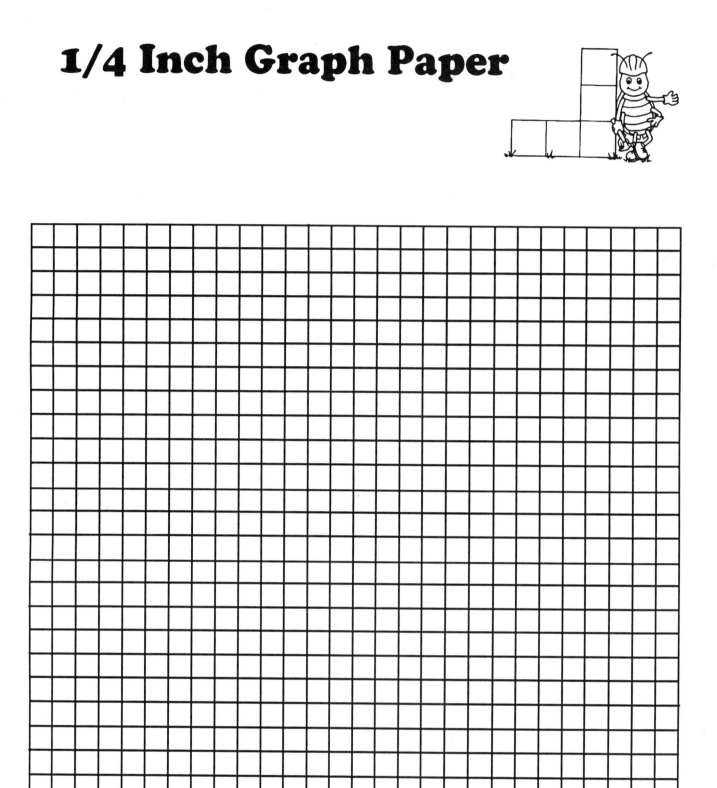

Problem Solving with Pentominoes
© 1992 Learning Resources, Inc.

Progress Chart

Name _____

Exploring Pentominoes

- ☐ Pentomino Shapes
- ☐ More Pentomino Shapes
- ☐ Find the Pentominoes
- ☐ Find More Pentominoes

Comments:

Moves and Patterns

- ☐ Slides
- ☐ Turns
- ☐ Slide or Turn?
- ☐ Flips
- ☐ Which Piece Matches?
- ☐ Cover Some Shapes
- ☐ Sliding Patterns
- ☐ Turning Patterns

Comments:

Covering Geometric Shapes

- ☐ Covering Shapes With Two Pieces
- ☐ Covering Shapes With Three Pieces
- ☐ Covering Shapes With Four Pieces
- ☐ Numbers
- ☐ A Shape to Cover
- ☐ Where Do the Pieces Fit?
- ☐ Turn or Flip to Fit
- ☐ Place the First Piece
- ☐ More Space to Cover
- ☐ Which Pieces?

Comments:

Puzzles and Games

- ☐ Two of the Same
- ☐ A Shape Puzzle
- ☐ Cover the Stairs
- ☐ Making Rectangles
- ☐ Making Large Rectangles

- ☐ A Dog
- ☐ A Butterfly
- ☐ A Flag
- ☐ Making Squares
- ☐ A Square Game

Comments:

Symmetry

- ☐ Symmetry of Pentomino Pieces 1 & 2
- ☐ Symmetry of a Shape
- ☐ Turn Symmetry
- ☐ More Turn Symmetry

Comments:

Area and Perimeter

- ☐ Perimeter of Pentomino Pieces
- ☐ Perimeter of Shapes
- ☐ Perimeter of a Rectangle
- ☐ Make a Figure
- ☐ Area of Shapes
- ☐ Area of a Square

Comments:

Congruence and Similarity

- ☐ Congruent Shapes
- ☐ Making Congruent Shapes
- ☐ Similar Shapes
- ☐ Doubling Pentominoes
- ☐ Tripling Pentominoes

Comments:

Pentomino Challenges

- ☐ Use a Small Model
- ☐ Animal Shapes
- ☐ Making a Box

Comments:

PENTOMINOES
GOOD WORK AWARD

To: _____

For: _____

_____
Teacher

_____
Date